GIScience and Geo-environmental Modelling

The "GIScience and Geo-environmental Modelling" book series seeks to publish a broad portfolio of scientific books addressing the interface between geography and the environment. The aim of the book series is to present geospatial technology approaches to data mining techniques, data analytics, modeling, risk assessment, visualization, and management strategies. The series includes peer-reviewed monographs, edited volumes, textbooks, and conference proceedings. The focus of Geo-environmental is on geospatial modelling in the frontier area of earth-environment related fields, such as urban and peri-urban environmental issues, ecology, hydrology, basin management, geohazards, estuarine-ecology, groundwater, agriculture, climate change, land-water, and forest resources, and related topics.

Geo-environmental modelling techniques have enjoyed an overwhelming interest in recent decades among the earth environmental and social sciences research communities for their powerful ability to solve and understand various complex problems and develop novel approaches toward sustainable earth and human society. Geo-environmental modelling using data mining, machine learning, and cloud computing technology is focused on spatiotemporal data analysis and modeling for sustainability in our environment.

Mrinmay Mandal ·
Nilanjana Das Chatterjee

Geo-Spatial Analysis of Forest Landscape for Wildlife Management

Springer

Mrinmay Mandal
Shyamsundarpur Junior High School
Paschim Medinipur, West Bengal, India

Nilanjana Das Chatterjee
Department of Geography
Vidyasagar University
Midnapore, West Bengal, India

ISSN 2730-7506 ISSN 2730-7514 (electronic)
GIScience and Geo-environmental Modelling
ISBN 978-3-031-33605-8 ISBN 978-3-031-33606-5 (eBook)
https://doi.org/10.1007/978-3-031-33606-5

This Springer imprint is published by the registered company Springer Nature Switzerland AG
The registered company address is: Gewerbestrasse 11, 6330 Cham, Switzerland

To my School Teachers

Naipur Santi Sudha Institution

Purba Medinipur, West Bengal, India

Foreword

Landscape level approach to biodiversity conservation is the norm of the day. Research in the field of conservation biology in last few decades has advanced our understanding on how a species interact with their environment and how different metapopulations interact with each other at a spatial and temporal level. The results indicate that long-term conservation of biodiversity requires an interconnected system of reserves/forest patches spread over an area large enough to offset ecological changes over time.

However, there is competing demand of landuse in a landscape. Often, legal boundaries further overlay on this landuse mosaic. This creates a fragmented landscape, where artificial barriers along legal ownership pattern create hindrance to gene flow. The unenviable task before the wildlife manager is to manage the species, sustainably, across the landscape despite having competing landuse demand, legal ownership issues and multiple resource demand from forest patches.

In very few areas of the world, the challenge of landscape level conservation is more onerous than in South West Bengal. The list of landscape attributes reads like a checklist of ideal problem landscape : highly fragmented forest with long forest edges, multiple use forest with full access to fringe-area-dwellers, agriculturally productive area producing three crops a year, high population density with more than 550 people living per square kilometre, existence of large number of linear infrastructure such as national highways, important railway lines, irrigation canals across the landscape, large-scale occurrences of human–wildlife conflicts and importantly presence of elephants, a mega herbivore in landscape. Very aptly, the authors of this book have chosen this challenging landscape to illustrate how wildlife management could be done here using a reference species of elephants.

The book starts with an introduction to landscape level management and then proceeds to elucidate the ecological concepts of wildlife habitat such as habitat shape, edge and connectivity. Important ecological tools such as Normalized Difference Vegetation Index (NDVI), Mean Nearest Neighbourhood Distance (MND), Mean Proximity Index (MPI) used in landscape level analysis have been introduced early in the book. The book illustrates use of Geospatial analysis tools for wildlife conflict area demarcation using Bankura District, West Bengal and Dalma Wildlife Sanctuary, Jharkhand as case studies. Using elephant as reference species, the book explains in sufficient details how species-specific corridor demarcation and wildlife conflict area demarcation can be done. The book concludes with practical suggestions on landscape improvement strategies and human–wildlife conservation strategies. A strength of the book is its logical progression: each chapter builds on what is discussed before and this gives a much needed coherence to the subject.

I believe the book would be very useful to serious wildlife enthusiasts and professional wildlifers alike. It would also serve as an advanced reference book for graduate students.

Professor Nilanjana Das Chatterjee and Dr. Mrinmay Mandal have worked and published extensively in the field of landscape level wildlife management, geospatial analysis of fragmented forest landscape, landuse alteration strategy, habitat suitability assessment and human–wildlife conflict, in the South West Bengal. In a sense, the book is a distillation product of their previously published work and decades of field experience.

I am confident that the labour of love that the book is, will be appreciated by the vast community of wildlife lovers.

Debal Ray
Principal Chief Conservator of Forests
Wildlife and Chief Wildlife Warden
Kolkata, West Bengal, India

Preface

The impact of the advancement of human civilization on the environment and biodiversity has been profound and detrimental. With the advancement of human civilization, far-reaching effects on the natural world started in the form of over-exploitation of natural resources, climate change, land use alteration and change, fragmentation of natural habitat, deforestation, habitat destruction, habitat loss, and so on. On the other hand, one of the indicators of such development is urbanization which encroaches on forest lands, and the establishment of settlements and agricultural land at the cost of natural ecosystems has a long-term impact on the natural environment. It is important to note that human civilization has made many positive contributions to biodiversity and the environment, such as conservation efforts, the establishment of protected areas, and the development of sustainable practices. However, the overall impact has been largely negative, and addressing these challenges is crucial for long-term sustainable ecosystems. These far-reaching consequences, including loss of biodiversity, altered landscapes, and habitat fragmentation increase the probability of movement of animal species to the fringe areas and result in human-animal conflicts.

Worldwide human and wildlife conflict is a result of such environmental degradation. Regionally, the wild animal species that still survive are struggling for their basic biological needs like food and safe shelter. The defeated species become vulnerable and are disappearing from this world. There is a massive loss in biodiversity that causes imbalance in the system, because this wildlife plays a huge role in the ecosystem. This imbalance ecosystem further creating various obstacles to sustain civilization and its proper development. Conserving wildlife not only protects the incredible diversity of life on Earth but also contributes to ecosystem stability, resilience, and the overall health of earth. It is an integral part of achieving a sustainable environment for the future. For this reason, various administrations, researchers, institutions and individuals around the world are working together to solve those problems and to save the wildlife. Wildlife conservation management is one of those efforts in which geospatial information and its analysis become a key role in present era.

Geospatial information is a very important for wildlife management research. Especially landscape information keeps strong influence on it because ecological activities of wild species like movement, colonization, connectivity, extinction, etc., are directly connected with the landscape at all scale. Wildlife interaction between landscape, i.e. habitat is very highlighted and important for wildlife conservation and conflict management. Habitat characters like structural configuration, amount, orientation, composition, etc., regulate wildlife movement and colonization as well as human–animal conflict due to their inter-patch movement and habitat preference activities. The present book introduces a new branch, i.e. landscape ecological model that is much cognizable in the present day context for wildlife management.

Currently, human–wildlife conflict (HWC) becomes a national issue in and in some cases international issue also, though its nature and characters are area specific. Most of the States of India face several HWC but human–elephant conflict (HEC) in different states of India are enormous in nature. In the State of West Bengal, HWC has diversified impact both for human society and wildlife communities like elephant, Royal Bengal tiger, crocodile, deer, wild boar, etc. In the state, HEC events attracts the most attention due to its high frequency, changing nature of conflict and massive societal loss. Considering these issues, the present book is written to realize the overall wildlife situation in the State of West Bengal. Regarding the prospect of the book it may be said that there are several books published all over the world. India has some specific books on same issues. But there are few books written to manage regional geospatial characters for specific wildlife management in the state. Specific wildlife management models and methods are applied using case study in the specific chapters of the present book, the major part of management models deal with land conversion and landuse planning which will decrease HWC and as well as HEC. These models are also applicable in other parts of the globe for wildlife management. Finally, in the last chapter some awareness generating steps are presented to conserve wildlife and minimize conflict events.

Egra, India Mrinmay Mandal
Midnapore, India Nilanjana Das Chatterjee

Contents

Abbreviations

AHP	Analytic Hierarchy Process
AWMPFD	Area Weighted Mean Patch Fractal Dimension
AWMSI	Area Weighted Mean Shape Index
BTR	Buxa Tiger Reserve
CA	Core area
CAD	Core area density
CD	Canal density
DBH	Diameter at Breast Height
DEM	Digital Elevation Model
DFO	District forest officer
DWS	Dalma Wildlife Sanctuary
ED	Edge density
FID	Field identity
HEC	Human–elephant conflict
HSI	Habitat suitability index
HWC	Human–wildlife conflict
IBT	Island Biogeography Theory
IUCN	International Union for Conservation of Nature
JFM	Joint Forest Management
LCP	Least Cost Path
LPI	Largest patch index
LULC	Landuse land cover
MCA	Mean core area
MNN	Mean Nearest Neighbour Distance
MPI	Mean Proximity Index
MPS	Mean Patch Size
MSI	Mean shape index
NC	Number of core
NH	National Highway
NND	Nearest neighbour distance
NTCA	National Tiger Conservation Authority
NUMP	Number of patch
PA	Patch area
PI	Proximity index
PLAND	Percentage of landscape
RD	Road density

SH	State Highway
TCA	Total core area
TCAI	Total core area index
TE	Total edge
WLS	Wildlife Sanctuary
WPD	Water patch density

List of Figures

List of Tables

List of Plates

List of Boxes

1 Introduction

Abstract

Landscape Planning and Wildlife management is an essential issue for biodiversity conservation in the present-day context. Ecology in recent days is augmented by spatial information science. Therefore, space or landscape becomes a new frontier of this subject. Landscape ecology has emerged as a discipline, focusing on spatial structure and landscape systems to understand regional ecological patterns and processes. The present study follows the same principle with the help of several branches of science, for addressing environmental issues and to understand ecology with the help of related subjects like geography, botany, zoology and surface geometry. Ecological distributions depend on the complex relationship between life communities and their environment in a specific landscape. Spatial character has been introduced with the horizontal and vertical function of ecosystem ecology. The present research investigates how spatial character, i.e. landscape surface pattern of the study area is related to species distribution because space is an important component to determine the diversity of life forms. The study makes a deep investigation to realize the relation between surface pattern (patch, corridor and matrix) and species behaviour. Landscape characters are very important in controlling ecological activities like animal movement and colonization. These processes are linked with habitat preference of individual species. The common and important word in ecology is 'Habitat'. It is directly related to the landscape. Quality of habitat depends on landuse and land cover attributes like patch shape, size, core, number and distance between patches. Analysis of these landscape characters and its relation with wildlife movement and colonization to understand and manage wildlife situations is the prime focus of this book.

1.1 Wildlife of the Study Area at a Glance

Wildlife resources constitute a vital link in the survival of the human species. Today, the effective conservation of wild animals is of great significance when wildlife habitats are under severe pressure and a large number of species of wild fauna having become endangered (Taber et al. 2003; Hundal 2004; Wilkening et al. 2019). India is the seventh-largest country in the globe with a long coastline of 7516 km. A great wealth of biological diversity exists in India's wetlands and marine areas. In a broad way, the variety and variability of life are present in the wildlife which has an enormous ecological value. Ecological balance is a state of dynamic equilibrium within a community of organisms in which genetic, species and ecosystem diversity remain relatively

M. Mandal and N. Das Chatterjee, *Geo-Spatial Analysis of Forest Landscape for Wildlife Management*, GIScience and Geo-environmental Modelling, https://doi.org/10.1007/978-3-031-33606-5_1

stable. The wildlife can be used to assess the quality of the environment. Some species sometime act as 'indicator species' for the health status of the ecosystem like Royal Bengal tiger in Sundarban (Sievers et al. 2020). This is why all predators are situated at the top of the trophic pyramid which can highlight environmental problems that occur at the lower levels, such as poisoning, pollution and disease (Chardonnet et al. 2002).

India is the house for about 350 species of mammals, 1224 species of birds, 408 species of reptiles, 197 species of amphibians and more than 2546 species of fishes according to the report of Wildlife Institute of India 2018–2019. As of now, 172 species of animals that are globally threatened, according to the IUCN lists, are found in India. This includes around 53 species of mammals, 69 species of birds, two species of reptiles and three species of amphibians. India has a number of Biodiversity Hotspot Zones that serves as a habitat to a number of wildlife species (Hundal 2004; Venkataraman et al. 2018). Among all the mammals, elephant the National Heritage Animal of India is endangered species in IUCN list. In spite of in the State of West Bengal, their population and habitation both are equivalently increased. Not only elephant but other wildlife diversity still exists depending upon increasing forest areas in the state.

According to the West Bengal State Forest Report 2019–20, state has a recorded forest area of 11,879 km^2. The forested areas are classified with a well-defined extent: (i) reserved forest that has an extent of 70.45 km^2, (ii) protected forests having an extent of 3772 km^2 and (iii) unclassed state forests with an area of 1053 km^2. The latest annual report 2019–2020 forest cover of West Bengal is 21.39% out of its total geographical area; this includes notified as well as forests created by afforestation outside the notified areas.

West Bengal has 4706 km^2 of forests under the protected network which is 39.50% of the total state's forest area and 5.28% of the total geographical area (West Bengal Forest Report 2019–2020). As the West Sundarban Wildlife Sanctuary has been declared as a new Wildlife Sanctuary, and the area of the Neora Valley National Park has increased, the protected forest areas now include as 1502.82 km^2. Two Tiger Reserves Sundarban and Buxa Tiger Reserve have in the state. There are two Elephant Reserves in the state also. The Mayurjharna Elephant Reserve covers an area of 414 km^2 and Eastern Dooars Elephant Reserve covers area of 977.51 km^2. State has also one Biosphere Reserve that is called the Sundarban Biosphere Reserve covering area of 9630 km^2. It includes Sunderban Tiger Reserve, core areas of Sunderban National Park, Halliday Island and Lothian Island Wildlife Sanctuaries with Sajnakhali Wildlife Sanctuary forming its buffer area.

From ancient period, West Bengal has rich species diversity in its wildlife (Jana et al. 2006). The state emphasizes in its conservation and management to achieve long-term biodiversity conservation. Several plans have been made to progress the management of the protected areas, reduce man animal conflict, develop the infrastructure, introduce habitat improvement programs and encourage involvement of local people in management of the protected areas (Alfred et al. 2012).

The glorious wildlife background of the state emphasizes in its conservation strategies, plans and projects. The curricular activities to conserve wildlife the state have several conserving units such like National Parks, Wildlife Sanctuary, specific species reserve, and biosphere reserves. These are listed below (Table 1.1).

1.2 Present Situation of Wildlife

Since industrial revolution, the state has lost its glorious wildlife background. Huge population pressure on natural land, social infrastructural development, frequent mining activities inside the forest and industrial establishment are responsible for becoming species endangered in nature as well as biodiversity loss and unexpected conflict between wildlife and human society. Human–animal conflicts now become a jumbo effect in West Bengal (Chatterjee 2016; Mandal and Chatterjee 2020b). North, south-

Table 1.1 District wise wildlife conservation units under West Bengal State

Name of the National Park	Area (km^2)	District
Neora Valley National Park	159.89	Kalimpong
Buxa National Park	117.10	Alipurduar
Gorumara National Park	79.45	Jalpaiguri
Sundarbans National Park	1330.10	South 24 Parganas
Jaldapara National Park	216.34	Alipurduar
Singalila National Park	78.60	Darjeeling
Name of the Wildlife Sanctuary	**Area (km^2)**	**District**
Chapramari WLS	9.6	Jalpaiguri
Senchal WLS	0.04	Darjeeling
Jorepokhri WLS	38.88	Darjeeling
Mahananda WLS	158.04	Darjeeling and partly Jalpaiguri
Bethudahari WLS	0.6686	Nadia
Ballavpur WLS	2.021	Birbhum
Ramnabagan WLS	0.145	Bardhhaman
Bibhutibhusan WLS	0.64	North 24 Parganas
Sajnakhali WLS	362.4	South 24 Parganas
Haliday Island WLS	5.95	South 24 Parganas
Lothian Island WLS	38	South 24 Parganas
Buxa WLS	314.52	Alipurduar
West Sundarbans WLS	556.45	South 24 Parganas
Name of the Tiger Reserves	**Area**	**District**
Buxa Tiger Reserve	390.5813	Alipurduar
Sundarbans Tiger Reserve	3345.7599	South and North 24 Parganas
Name of the Biosphere Reserve	**Area (km^2)**	**District**
Sundarbans Biosphere Reserve	9630	South and North 24 Parganas
Name of the Elephant Reserves	**Area (km^2)**	**District**
Eastern Dooars Elephant Reserve	977.51	Jalpaiguri and Alipurduar
Mayurjharna Elephant Reserve	414	Jhargram, Bankura and Purulia

western and Sundarbans region of the state, comes in front of several problems like unexpected movement of wildlife, food crisis, unethical anthropocentric intrusion into the forest and finally loss of life and assets (Islam et al. 2009; Vyas 2012). HWC intensity and its impact become the prime issue worldwide (Distefano 2005) as well as in the state (Mukherjee 2016). For that purpose, state and central agencies, field ecologists, researchers and stakeholders worked hard to prevent this situation. Both state and central bodies arrange some programs to understand the overall condition and manage the diverse effect of HWC. One step is census program which is taken up in various years for various animals. This is done in order to keep in check the population of endangered species. This also provides us the insight knowledge how organisms interact with each other and with their environments. It helps to understand the reasons of population growth or shrink of that particular animal and helps to better predict the future changes.

Frequent census program for the purpose of the wildlife is taken up; this also ensures a proper understanding of the historical as well as present situation of wildlife. When the population of a certain wildlife comes to a better number, it helps

Table 1.2 Recent animal census that had been taken up by the state forest department

Range/Unit—Sundarban		No. of individuals by year				
		2016–17	2018–19	2019–20		
Sajnekhali Wildlife Sanctuary	Tiger	14	15	10		
Basirhat Range		14	13	19		
National Park (West) Range		19	18	20		
National Park (East) Range		16	20	24		
24 Parganas (South) Division		24	22	23		
Year	Elephant	2010	2014	2017		
Elephant		529 (NB) 118 (SB)	590	488 (NB) 194 (SB)	NB-North Bengal SB-South Bengal	
Year Sundarban Tiger Reserve and South 24 Parganas Division	Crocodile	Indirect sighting	Direct sighting			
			Adult	Juvenile	Hatchling	Total
2012		99	69	61	10	140

Source State Annual Report 2019–2020

to conserve the ecosystem and a proper way of preservation of the forests can be done (Table 1.2).

1.3 Wildlife Conflict-Related Issues

Human–wildlife conflict (HWC) arises "when wildlife's basic biological requirements extend beyond the human populations areas mentioned by World Conservation Union. Contact with wildlife occurs in both the region of suburban and rural areas directly or indirectly. Generally, it is more frequent inside and around the protected areas (Mandal et al. 2021b). May be the contact zone was under wildlife habitat. Later these lands are claimed by human society. As a result, wildlife habitat becomes thinner. The number of wildlife population increases or becomes vulnerable due to disturbances such as insufficient food supply. As a result they often come closer to adjacent cultivated fields or grazing areas (Distefano 2005)

Similar circumstances have been found in some pockets in West Bengal (Fig. 1.1). Such as Human–elephant conflict (HEC) is become jumbo issue in northern districts like Jalpaiguri, Alipurduar (Anwar et al. 2015; Mukherjee 2016) south-western districts like Purulya, Bankura, Birbhum, Paschim Medinipur and Jhargram (Singh 2006; Mandal et al. 2015; Mandal et al. 2021a; Majumder 2022). Human–tiger and crocodile conflict is a popular incident in most southern part of the state, i.e. Sundarbans region. Also other conflict events related to wild animals like rhino, leopard, spotted deer, Gaur-India bison, etc., happened in regular basis in the same part of the state. So the increased intensity of conflict and number make a vital issue all over the society and nation. Wildlife conflict makes not only loss of life and assets, its function gradually propagates through society, economy and ecology in broad sense step by steps (Dickman 2010).

1.3.1 Social Issues

Human activities and sensations are the primary driver of biodiversity loss and imbalance. It is the most important determinant factor of species persistence. The individual and community perception of nature (including wildlife) depends on the social context. There is big relation between advancement of civilization and wildlife. Becoming the most intelligent animal in the globe, human being use to collect interest from

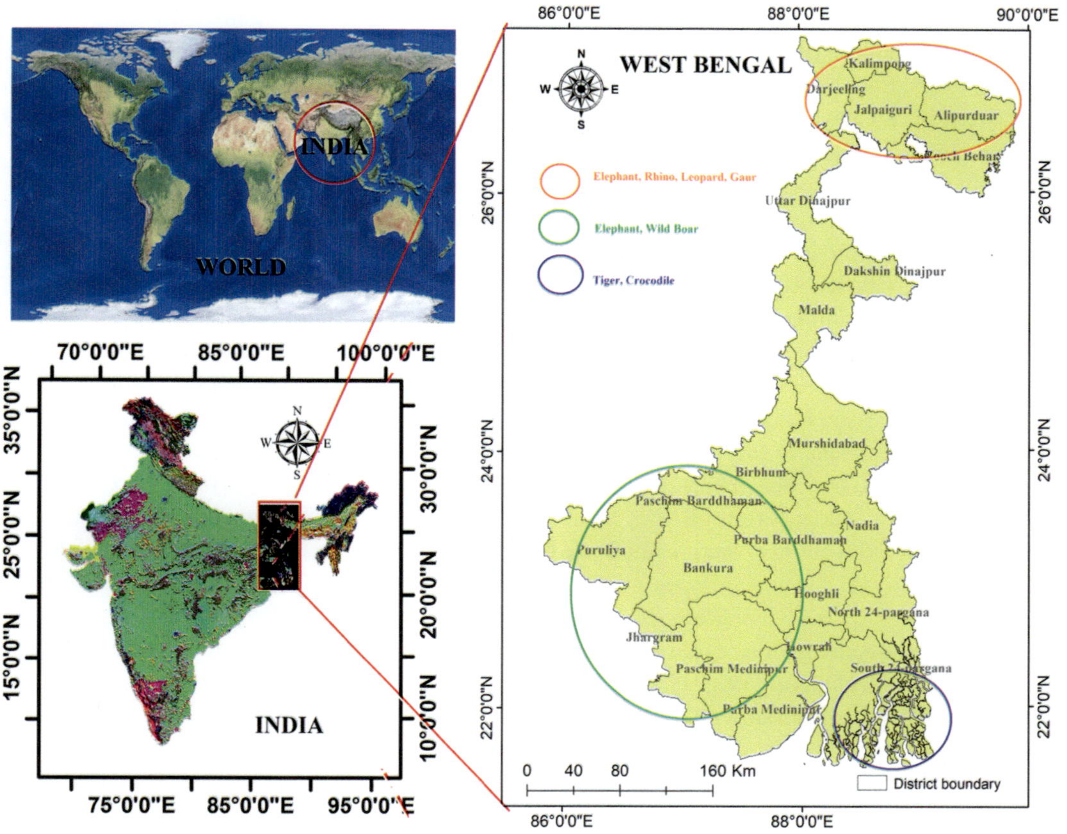

Fig. 1.1 Location of major wildlife conflict zones under West Bengal State with respect to national and international location

wildlife for their successive section of development. Wildlife provides some important components for human society from beginning of their origin (Chardonnet et al. 2002). Wildlife offers numbers of resource and supports the entire livelihood of the society stated by Manfredo et al. (2009). Very often, wildlife is also responsible for human casualties (accidents which wound or kill people), predation of domestic animals (Kala et al. 2013), depredation to crops (Mandal 2018) and destruction of houses and domestic assets (Sukumar 2003). It is true that any kind of depredation is not necessarily linked with its size and amount.

In South Bengal for instance, massive crop raiding occurs by the herds of elephant that migrate from Dalma Wildlife Sanctuary, Jharkhand State during certain period of the year. Another wild animal wild boar is also responsible for crop damage in certain places of South Bengal. Wildlife damage to agricultural crops is a serious concern affecting much of the world today. Decline in the availability and efficiency of natural prey/food sources lead to wild animals looking for alternate sources. Alternately, new resources created by humans are easily available and draw wildlife, resulting in conflict. Local communities' perception about wildlife inhabiting in and around their surroundings is a vital question for biodiversity conservation (Mekonen 2020). The problems of the local indigenous communities regarding wildlife if ignored, then community will be affected. Social factors should be taken into full consideration to promote human–wildlife coexistence. These understandings make a sense regarding human–wildlife conflict and human–human conflict on wildlife (Pimid et al. 2022).

Ignorance suppresses community's capability to handle situation appropriately if they receive the wrong conservation messages (Jacobsan et al. 2019). However, effective conservation messages encourage public support for conservation (Kidd et al. 2019). The local original community can be empowered through a socio-economic approach. Such approach is encouraging and giving equal opportunities to them to participate in wildlife-based activities. It will provide income and maintain the impact of tight conservation rules.

Another social issue is ex-gratia that is being paid by the government to the members of the affected family. It has become a partial but important step for damage control. The procedure and legal verification is very complicated and time taking to get the compensation. It makes controversy among the affected communities. Another issue is ignorance and slow steps against animal movement driving, chasing, rescuing and unhelpful behaviour with local victims or villagers though the departmental authority wants that the wildlife does not disturb the humans. Insufficient infrastructure and equipments to control wildlife indirectly raise the social problem (Johnson et al. 2018). Even sometimes, the forest department becomes disappointed with some rare unexpected condition.

1.3.2 Economic Issues

Wildlife is a source of both costs and benefits to society. Costs occur from wildlife predation on livestock, destruction of crops, traffic collisions, and transmission of diseases to animals and humans. Benefits accrue from hunting, recreational activities, food, and other ecosystem services. Approximately 50% of all mammals in the globe are in decline, and 25% are facing extinction. It occurs because of illegal hunting (Treves et al. 2003) and destruction of habitats due to agricultural purposes (Roemer and Forrest 1996; Woodroffe and Ginsberg 1998). However, wildlife also provides a source of recreational benefits and income like wild animal viewing can generate significant profits (Barnes et al. 1999; Hofer 2002). The economic importance of wildlife is as difficult to appraise in developing countries, but in developed countries it has a classic academic exercise.

In Sundarban, there are tourism packages in which visitor can spend nights in the creeks of jungles and enjoy the nature following strict guidelines. But guides denied it for extra earning and break the rules. Tourist excitement also flourished them. As a result, sometime visitors were attacked by wildlife and they did not follow the rules and regulation. The violence of rules make irritate wildlife like Tiger when they cross the river in Sundarbans (Vays 2012; Das 2015). The poor fisher man did not maintain the rules of Biosphere Reserve, and they penetrate into the core zone. Accident or even deaths of fisher man and honey collectors by wild attack are economically weaker the family in Sundarban. Several missings of human being are recorded especially the poor river duellers who was the only earning member of the family. Another opposite event also happened that wildlife especially tiger in Sundarban comes into the locality and killed cattle, cows and goats even child, that's why the poor families suffer from such kinds of incidents; thus, the wildlife attacks have a significant impact on economy of the region.

Similar wildlife event regularly occurred, but the nature of conflict is little bit different in North and South Bengal. Here elephant depredation is the key issue in local economy (Mandal et al. 2020a). Huge crop areas and vegetable gardens are damaged by elephant and wild boar. It makes intolerable economic pressure on affected family. Only ex-gratia and compensation given by the authority do not play an ultimate policy to recover the economic condition of the affected families. As a result, sometimes affected family and their neighbours burst into anger against concern authority. Another conflict is human death and injure. This accident now becomes incident in nature in South Bengal. News of human death or injure comes in front in regular basis. Every unexpected death of human is related to their family income.

Political Issues

Wildlife conservation has been a public issue since civilization. Much of the dispute over wildlife conservation involves property and property rights. Rather wildlife conflict is now being a political issue than social and economic (Margulies et al. 2018). The present local circumstance is now controlled by local privileged. Political penetration is duly generated when wildlife issue covers among one or more communities. In South Bengal, it is noticed that nearest two villages are ensued by the arguments related to elephant herds driving and chasing. Sometimes these arguments ended with physical conflict between them and converted to a political issue. Similarly when a wild animal such as elephant was died by electric fencing or other human intervention, then political issue also raised by local community. Political issue massively stands up during failure or ignorance of concern wildlife authority for the purpose of driving, rescuing, translocation wild animal for the demand of local community.

Ecological Issues

One of the most wildlife-related ecological issues is HWC. Nature of HWC determines other discussed secondary issues like social, economic and political. HWC has several natural background like climate change, tectonic movement, volcanic activities, flood, drought, desertification, etc., but a major part of HWC is covering with landscape characters stated by Teixeira et al. (2021). Moreover, overlap of both human and wildlife domain is the finest cause of it. Therefore, landscape diversity and biodiversity assimilate a series of process by which regional ecosystem be balanced (Messmer 2009). It is true and interesting that landscape has a capacity to modify the function of biodiversity but biodiversity has not reciprocally the same. So the landscape characters, composition, variation and structural orientation regulate wildlife ecological pattern and process such as movement, colonization, habitation, extinction and predation. These unusual patterns and processes make HWC argued by several scholars (Forman 1995; Mandal et al. 2018, 2019) and also created by habitat fragmentation, habitat loss, reformation of newly habitat, human corridor establishment along with wildlife natural movement path, appearance of secondary food efficiency and its seasonal variation, water crisis, disturbance and even unscientific habitat modification. So these are the ecological issues related to HWC.

From upper discussion, it is evident that landscape operates a significant ecological function for HWC. Discussed all ecological issues have great extant in the State of West Bengal. Forest or habitat fragmentation and loss of habitat are the most important consideration for understanding the wildlife circumstance in this state.

1.4 Spatio-temporal Changes of Landuse in West Bengal

Geographically, the landuse in West Bengal comprises with some common important classification (Fig. 1.2). Historically, huge land alteration occurred due to maximum revenue collection under the late British rule in Bengal province. But after independence, the momentum goes faster for controlling population pressure and mitigates basic needs like food and shelter. As a result, a radical change was found in the land of agriculture and forest (Best and Champion 1970) which influenced the regional ecological behaviour as well as the whole existing biodiversity. After implementation of landuse planning, unproductive lands become productive in West Bengal. Net sown area and measurable percentage of forest area gradually increased at the cost of barren and fallow land (Table 1.3).

1.4.1 Forest Landscape

An attractive diversity is originated in the forest cover of West Bengal. The nature of forest is

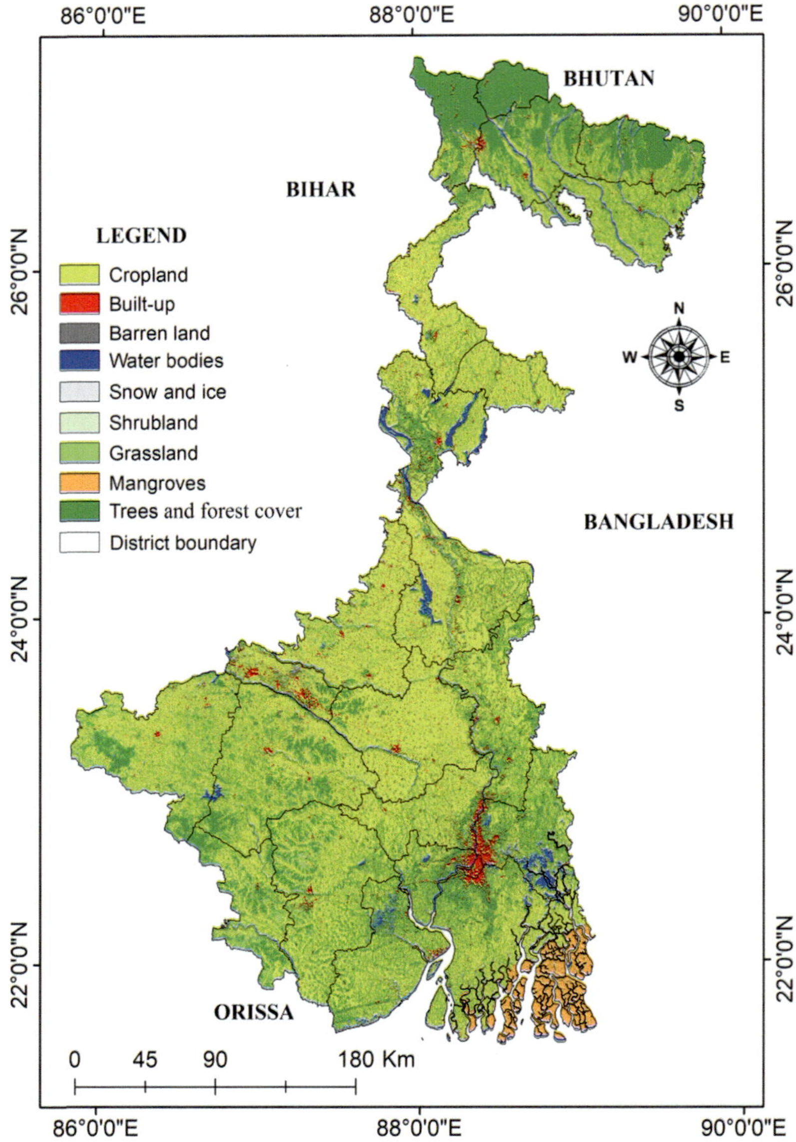

Fig. 1.2 Landuse/Landcover classification map of West Bengal, 2020. *Data Source* ESA World Cover, 2020

diverse due to climatic, topographic and soil diversity in the state. On the basis of these factors, the state forest is categorized into broad four sections. These are subalpine and alpine forest in the district of Darjeeling, tropical moist deciduous, tropical dry deciduous and littoral swamp in Sundarban (Fig. 1.3 and Plate 1.1). Another classification of forest, based on forest land of which 7054 km^2 is reserved forest, 3772 km^2 is protected forest and 1053 km^2 is unclassed forests. The total recorded forest area is 11,879 km^2. As per the latest report published by Forest Survey of India, the forest and tree cover of the state is 21.39% of the geographical area

Table 1.3 Temporal changes of landuse from 2003 to 2018 under West Bengal

Landuse type	2003–2004	2008–2009	2017–2018
Geographical area	8875	8875	8875
Reporting area for land utilization	8687.54	8684	8684
Forest area	1171.31	1174	1175
Not available for land cultivation	1608.97	1793	1873
Barren and unculturable land	27.04	21	9
Permanent pastures and other grazing land	5.05	7	2
Land under mics, tree cops and groves	57.87	55	48
Culturable Wasteland	34.50	32	14
Fallow land other than current Fallows	22.31	22	7
Current fallows	333.38	287	308
Net area sown	5463.67	5294	5247

Area in thousand hectares

Source Directorate of Economics and Statistics, Ministry of Agriculture and Farmers Welfare, Government of India

which is 88,752 km^2. Forest cover without tree cover of the state is 13.38% as per State Annual Administrative Report 2019–20.

From late decades of twentieth century, West Bengal forest areas gradually enrich its quantity and quality (Fig. 1.4). This enormous fact is also verified by temporal image analysis (Fig. 1.5). Basically, well planned approach as such Joint Forest Management (JFM) and some aforestation program implementation has been enhancing forest quantity in the recent past (Pattnaik et al. 1997; Banerjee 2007). As a result, wildlife has come back in some pockets of forested areas under this state (Jana et al. 2014; Ahmed et al. 2017). But it is unfortunate to see that Sundarban Biosphere suffering from casualty of forest loss.

District wise forest cover scenario depicts that South 24 Parganas occupy the highest cover, i.e. 42.38% and consequently Bankura 20.81%, Paschim Medinipur 14.41% and Jhargram 28.65% in southern district of the state. In North Bengal Darjeeling, Jalpaiguri, Kalimpong and Alipurduar districts hold larger amount of forest cover respect other district (Table 1.4).

The districts of West Bengal hold distinct and diverse wildlife habitat based on the rich forest cover. Not only forest cover but also other factors like climate, tidal activities, geomorphic features, etc., are responsible for forming the ecosystem diversity as well floral and faunal communities in the state. Sundarban Biosphere in South 24 Parganas is an example for factor responsibility. Here tidal activities and nature of mangroves make an unique ecosystem which is far different from other terrestrial ecosystems in different districts.

1.4.2 Non-forest Landscape

The forest landscape holds major part for performing biological function of wildlife. But other corresponding landuse also play significant role. Forest gives the basic shelter, resting ground and forage for some herbivores animal, but the presence of other landuse inside or edge of the forest helps wildlife to live their life. Water body inside or outside of the habitat is an essential example. In opposite word, over exploitation of

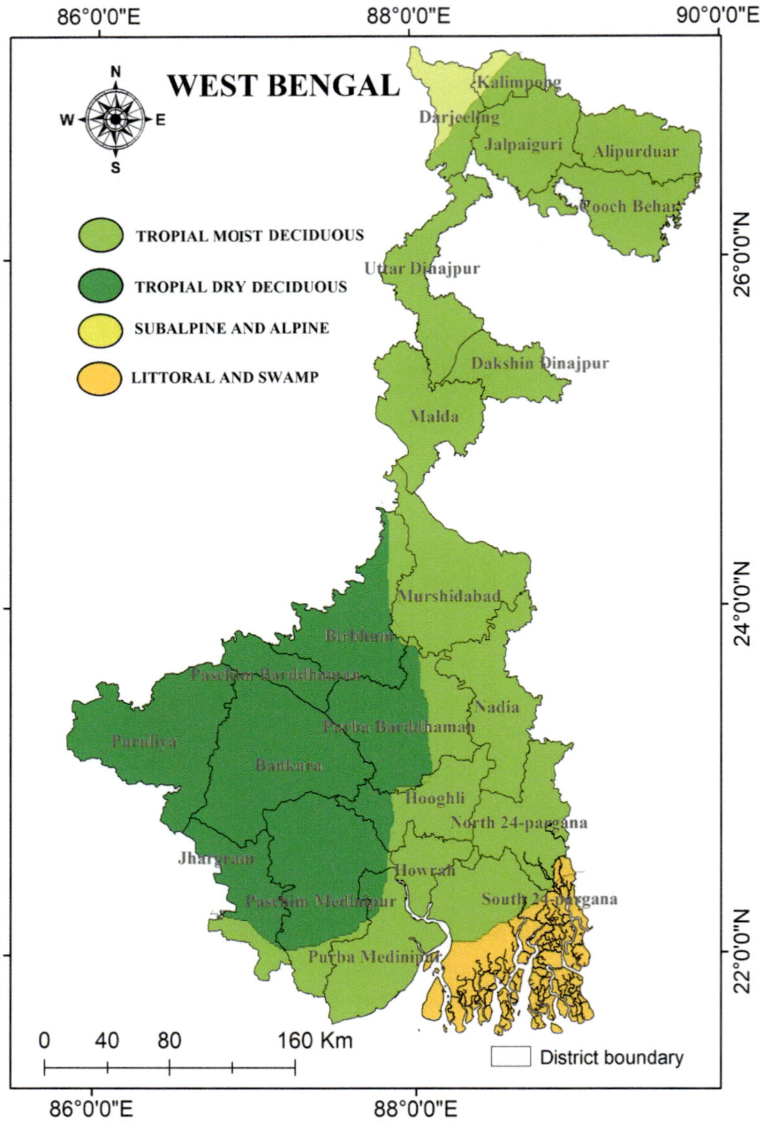

Fig. 1.3 Broad categories of natural forest in West Bengal

other landuse such as road network, industrial unit, built-up areas and human modified landuse gives harmful effect on wildlife ecosystem (Farina 2008). So the intensity and proper planning of landuse have remarkable for wildlife management.

In West Bengal, unscientific or improper landuse affects ecosystem diversity (Mandal et al. 2021c) in the districts where wildlife activities are in function. Settlements, road network, (Plates 1.2 and 1.3) area of crop lands at the cost of barren or vacant land has been recorded State Annual Report and Directorate of Economics & Statistics. These may be important indicator for human development index, but it encourages wildlife conflict. Again proportional wetlands areas become declined which has negative impact on ecosystem. Rapid urbanization and socio-economic development in the vicinity of forest land and wetlands in National Park, Wildlife Sanctuary and Biosphere areas damage and create disturbance in ecosystem.

Plate 1.1 **a** Tropical moist deciduous in Gorumara National Park-Dooars, **b** tropical dry deciduous forest in Joypur forest under Panchet forest division, Bankura district and **c** Mangrove in Indian Sundarbans

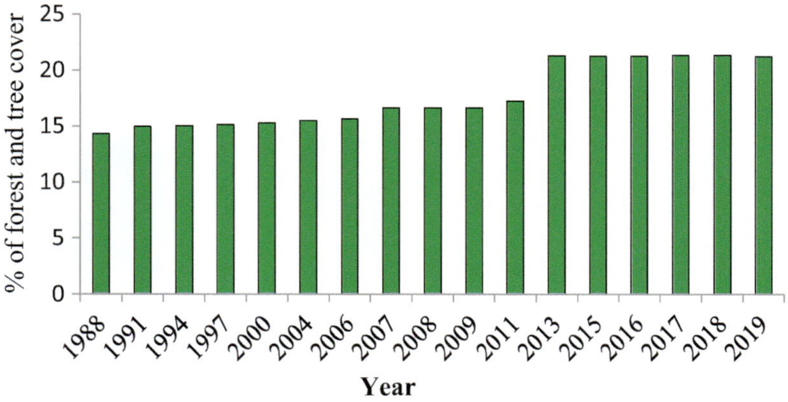

Fig. 1.4 Temporal variation of forest and tree cover percentage under State of West Bengal. *Source* State Annual Forest Report, 2019–2020

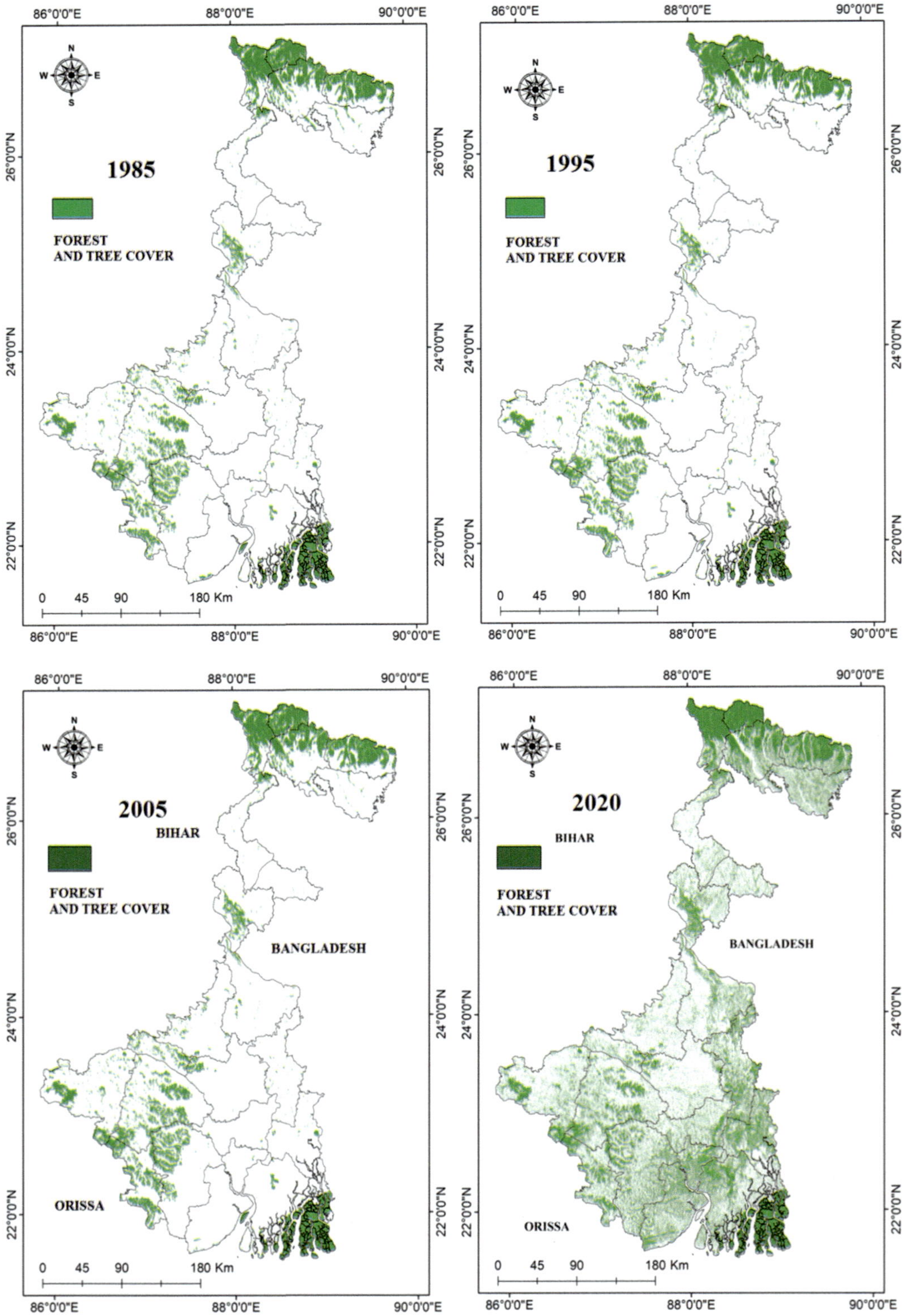

Fig. 1.5 Temporal forest and tree cover change detection maps of West Bengal from 1985 to 2020

Table 1.4 District wise forest and tree cover in % with respect to geographical area under State of West Bengal

S. No.	District	Geographical Area in (km^2)	Recorded Forest Area in (km^2)	Percentage (%)
1	Alipurduar	3383	1067	31.54
2	Coochbehar	3387	36.01	1.06
3	Jalpaiguri	2844	723	25.42
4	Darjeeling	2092.5	834.3	39.87
5	Kalimpong	1044	369.7	35.41
6	Uttar Dinajpur	3140	6.02	0.19
7	Dakhin Dinajpur	2219	8.27	0.37
8	Malda	3733	16.87	0.45
9	Murshidabad	5324	7.7	0.14
10	Nadia	3927	12.33	0.31
11	Bhirbhum	4545	159.87	3.52
12	Purba Bardhaman	5432.69	124.53	2.29
13	Paschim Bardhaman	1603.17	135.38	8.44
14	Bankura	6882	1431.92	20.81
15	Purulia	6259	922.91	14.75
16	Jhargram	3037.64	870.19	28.65
17	Paschim Medinipur	6308	908.81	14.41
18	Purva Medinipur	4736	18.84	0.4
19	Hoogli	3149	3.37	0.11
20	Howrah	1467	0	0
21	Kolkata	185	0	0
22	24 Parganas North	4094	0.96	0.02
23	24 Parganas South	9960	4221	42.38
Total		88,572	11,879	13.38

Source Annual Forest Report 2019–20

1.5 Significance of the Study

A big crisis and threats observed in recent era are related to imbalance of ecosystem. As a result government and government supported private agency stress the importance and understand the pressure on it. Therefore, different kinds of planning and management plans are made to address the adverse situation on ecosystem. Wildlife takes part a major role for stabilizing the ecosystem. So, wildlife conservation is the practice of protecting plant and animal species and their habitats. As part of the world's ecosystems, wildlife provides balance and stability to natural processes. The goal of wildlife conservation has two dimensions; one is to ensure the survival of these species, and another is to educate people on living sustainably with the species. This work helps to people of all ages, and it appreciates our country's natural resources. Further it helps to understand how to conserve those resources for future generations. Through structured

Plate 1.2 Different human modified landuse in Bakura district under West Bengal. **a** Railway line near Peardoba under Bankadaha forest range, **b** state highway-2 across through the Joypur forest, **c** a guest house near Susunia hill and **d** an agricultural landscape in the foothill of Susunia

educational experiences and activities targeted to varying populations, conservation education that enables people to realize how natural resources and ecosystems affect each other and how resources can be used wisely.

Concentrating the issue, this chapter analysed the habitat composition and its ecological outcomes for understanding the overall situation. The analysis of data and information show the current status of the animals in their respective environments from the case studies. All investigation commonly has a centre of attention on landscape because approach of landuse planning is the base of constructing the pillar for reaching secondary ecosystem planning. Another reason is different behaviour of wildlife that are related to landscape diversity. Therefore, a comprehensive study has been done depending on those factors for proper mitigation, and it can bring up a result that will help in conserving the indigenous wildlife of the state. The chapter also explains the relationship between wildlife and humans and its effect by some case studies. By studying all the above aspects, it will be very convenient to

Plate 1.3 Forest edge agricultural practice in Bankadaha forest range under Bankura district

prepare a conservation and management plan for sustainable Wildlife Habitat.

The approaches, information, methods and results inside this book open further avenues for reading and research.

References

Ahmed MU, Jana SK (2017) Role of joint forest management (Jfm) in improving the living standard of rural people: some case studies in West Bengal in India. Int J Hum Resour Res Rev 5:16–26

Alfred JRB, Sanyal AK, Tiwari S, Mitra S (2012) Status of biodiversity of West Bengal. Final technical report, Department of Environment. Kolkata, West Bengal, India

Anon (2019–20) State forest report. Government of West Bengal. Directorate of forests office of the principal chief conservator of forests. Kolkata

Anwar R, Saralch HS, Kumar S (2015) Human–wildlife conflict: issues versus mitigation. Indian For 141 (12):1305–1314

Banerjee A (2007) Joint forest management in West Bengal. In: Forests, people and power: the political ecology of reform in South Asia, pp 221–260

Barnes JI, Schier C, Van Rooy G (1999) Tourists' willingness to pay for wildlife viewing and wildlife conservation in Namibia. S Afr J Wildl Res 29(4):101–111

Best RH, Champion AG (1970) Regional conversions of agricultural land to urban use in England and Wales, 1945–67. Transactions of the Institute of British Geographers, 15–32

Chardonnet P, Clers BD, Fischer J, Gerhold R, Jori F, Lamarque F (2002) The value of wildlife. Revue scientifique et technique-Office international des epizooties 21(1):15–52

Chatterjee ND (2016) Man-elephant conflict: a case study from forests in West Bengal India. Springer International Publishing

Das CS (2015) Causes, consequences and cost-benefit analysis of the conflicts caused by tiger straying incidents in Sundarban, India. In: Proceedings of the zoological society, vol 68, no. 2. Springer India, pp 120–130

Dickman AJ (2010) Complexities of conflict: the importance of considering social factors for effectively

resolving human–wildlife conflict. Anim Conserv 13 (5):458–466
Distefano E (2005) Human-Wildlife conflict worldwide: collection of case studies, analysis of management strategies and good practices. Food and Agricultural Organization of the United Nations (FAO), Sustainable Agriculture and Rural Development Initiative (SARDI), Rome, Italy. Available from: FAO Corporate Document repository. http://www.fao.org/documents
Farina A (2008) Principles and methods in landscape ecology: towards a science of the landscape, vol 3. Springer Science & Business Media
Forman RTT (1995) Land mosaic: the ecology of landscape and regions. Cambridge University Press, Cambridge, England
Hofer D, Blanco JC (2002) The Lion's share of the hunt: trophy hunting and conservation: a review of the legal Eurasian tourist hunting market and trophy trade under cites: a traffic Europe Regional Report. Traffic Europe
Hundal SS (2004) Wildlife conservation strategies and management in India: an overview. In: Proceedings of the species at risk 2004 pathways to recovery conference, Victoria conference centre, BC, Canada, pp 2–6
Islam SN, Gnauck A (2009) Threats to the Sundarbans mangrove wetland ecosystems from transboundary water allocation in the Ganges basin: a preliminary problem analysis. Int J Ecol Econ Stat 13(W09):64–78
Jacobson SK, Morales NA, Chen B, Soodeen R, Moulton MP, Jain E (2019) Love or loss: effective message framing to promote environmental conservation. Appl Environ Educ Commun 18(3):252–265
Jana G, Misra KK, Bhattacharya T (2006) Diversity of some insect fauna in industrial and non-industrial areas of West Bengal India. J Insect Conserv 10 (3):249–260
Johnson MF, Karanth KK, Weinthal E (2018) Compensation as a policy for mitigating human-wildlife conflict around four protected areas in Rajasthan India. Conserv Soc 16(3):305–319
Kala CP, Kothari KK (2013) Livestock predation by common leopard in Binsar Wildlife Sanctuary, India: human–wildlife conflicts and conservation issues. Hum-Wildl Interact 7(2):325–333
Kidd LR, Garrard GE, Bekessy SA, Mills M, Camilleri AR, Fidler F, Fielding KS, Gordon A, Gregg EA, Adams VM (2019) Messaging matters: a systematic review of the conservation messaging literature. Biol Conserv 236:92–99
Majumder R (2022) Human-elephant conflict in West Bengal, India: present status and mitigation measures. Eur J Wildl Res 68(3):1–14
Mandal M (2018) Forest range wise Asian Elephant's (Elephas Maximus) Habitat suitability assessment through Food and Water availability: a case study in Panchet Forest Division, Bankura, West Bengal
Mandal M, Chatterjee ND (2018) Quantification of habitat (forest) shape complexity through geo-spatial analysis: An ecological approach in Panchet forest division in Bankura, West Bengal. Asian J Environ Ecol 6:1–8
Mandal M, Chatterjee ND (2019) Forest core demarcation using geo-spatial techniques: a habitat management approach in Panchet Forest division, Bankura, West Bengal India. Asian J Geogr Res 2(2):1–8
Mandal M, Chatterjee ND (2020a) Elephant's habitat suitability assessment through geo spatial quantification in Panchet forest division, West Bengal. Ecofeminism Clim Change
Mandal M, Chattarjee ND (2020b) Geo-statistical analysis to understand nature of forest patch shape complexity in panchet forest division under Bankura district, West Bengal. Indian J Ecol 47(1):96–101
Mandal M, Chatterjee ND (2021a) Geospatial approach-based delineation of elephant habitat suitability zones and its consequence in Mayurjharna Elephant Reserve, India. Environ Dev Sustain 23(12):17788–17809
Mandal M, Chettarjee ND (2021b) Human-Elephant conflict in Joypur Forest Influence Areas, West Bengal, India. Gajah 54:34–36
Mandal M, Chatterjee ND (2021c) Spatial alteration of fragmented forest landscape for improving structural quality of habitat: a case study from Radhanagar Forest Range, Bankura District, West Bengal, India. Geol Ecol Landscapes 5(4):252–259
Mandal M, Chettarjee ND, Hazra J (2015) Elephant migration and colonization in Bankura district, West Bengal, India. Vidyasagar University. Indian J Geogr Environ 14:46–52
Manfredo MJ, Teel TL, Zinn H (2009) Understanding global values toward wildlife. Wildl Soc Sci Hum Dimensions 31–43
Margulies JD, Karanth KK (2018) The production of human-wildlife conflict: a political animal geography of encounter. Geoforum 95:153–164
Mekonen S (2020) Coexistence between human and wildlife: the nature, causes and mitigations of human wildlife conflict around Bale Mountains National Park Southeast Ethiopia. BMC Ecology 20(1):1–9
Messmer TA (2009) Human–wildlife conflicts: emerging challenges and opportunities. Hum-Wildl Conflicts 3 (1):10–17
Mukherjee N (2016) A brief appraisal of human wildlife conflict in Jalpaiguri and Alipurduar districts of West Bengal. Int J Sci Res Pub 6:131–136
Pattnaik BK, Dutta S (1997) JFM in South-west Bengal: a study in participatory development. Econ Polit Wkly 3225–3232
Pimid M, Mohd Nasir MR, Krishnan KT, Chambers GK, Ahmad AG, Perijin J (2022) Understanding social dimensions in wildlife conservation: multiple stakeholder views. Animals 12(7):811
Roemer DM, Forrest SC (1996) Prairie dog poisoning in northern Great Plains: an analysis of programs and policies. Environ Manage 20(3):349-359
Sievers M, Chowdhury MR, Adame MF, Bhadury P, Bhargava R, Buelow C, Friess DA, Ghosh A,

Hayes MA, Connolly RM (2020) Indian Sundarbans mangrove forest considered endangered under Red List of Ecosystems, but there is cause for optimism. Biol Conserv 251:108751

Singh AK (2006) Ecological investigation of human-elephant conflicts in South West Bengal, Doctoral dissertation, Saurashtra University

Sukumar R (2003) The living elephants: evolutionary ecology, behaviour and conservation. New York: Oxford University Press

Taber RD, Payne NF (2003) Wildlife, conservation, and human welfare: a United States and Canadian perspective. Malabar, Florida: Krieger Publishing Company

Teixeira L, Tisovec-Dufner KC, Marin GDL, Marchini S, Dorresteijn I, Pardini R (2021) Linking human and ecological components to understand human–wildlife conflicts across landscapes and species. Conserv Biol 35(1):285–296

Treves A, Karanth KU (2003) Human-carnivore conflict and perspectives on carnivore management worldwide. Conserv Biol 17(6):1491–1499

Venkataraman K, Sivaperuman C (2018) Biodiversity hotspots in India. In: Indian hotspots. Springer, Singapore, pp 1–27

Vyas P (2012) Biodiversity conservation in Indian Sundarban in the context of anthropogenic pressures and strategies for impact mitigation. Doctoral dissertation, Saurashtra University

Wilkening J, Pearson-Prestera W, Mungi NA, Bhattacharyya S (2019) Endangered species management and climate change: when habitat conservation becomes a moving target. Wildl Soc Bull 43(1):11–20

Woodroffe R, Ginsberg JR (1998) Edge effects and the extinction of populations inside protected areas. Science 280(5372):2126–2128

Characterizing Major Wildlife Habitats in West Bengal

2

Abstract

The State of West Bengal, in the Eastern region of India, is home to a rich and variety of forest habitat and wildlife. From the famous Royal Bengal tiger of Sundarbans, the crocodiles of Gangetic delta and the Spotted deer of Nadia, Birbhum, Burdwan and North 24 Parganas to the Wild boar of South Bengal, the diversity of south Bengal wild species are very rich. West Bengal, in India is becoming a new destination for elephants. There are two distinct wild elephants regions spread over North Bengal (Jalpaiguri and Darjeeling) and South Bengal (W. Midnapur, Jhargram Bankura and Purulia districts). Mainly four districts Bankura, Purulia and Paschim Medinipore and Jhargram in West Bengal facing severe problem of Human Elephant Conflict related to extended home range of elephant migrated from Dalma Wildlife Sanctuary, Jharkhand. The problems initiated after 1980 and the tendency of such events gradually increased in the consequent years. Both the number of migrated elephants and the duration of stay increased in the newly selected habitat. This event has raised the situation of man elephant conflict in South West Bengal. Present research investigates the factors behind such habitat colonization and or selection of wild animals with special reference to South Bengal Elephant as Keystone species through spatial and temporal analysis using Geospatial techniques.

2.1 Elephants in India: A Brief Introduction

Asian elephant (*Elephas maximus*) is the only mega-herbivore living in some areas of south East Asia (Menon and Tiwari 2019). These elephants (species genus-*Elephas*) in Asia categorized in four subspecies (Shoshani and Eisenberg 1982; Sukumar 1992, 2003). The Sri Lankan elephant (*Elephas maximus maximus*) is native to Sri Lanka, (*Elephas maximus indicus*) native from main land in Asia; the subspecies (*Elephas maximus sumatranus*) belong in the island of Sumatra (*Elephas maximus asurus*) have become extinct in ancient period (Choudhury 1980; Choudhury et al. 2008). These extinct subspecies of Asian elephant have been distributed once upon a time in the Middle East countries like Iran, Iraq, Syria and Turkey. They become disappeared from at least 1800 BC and likely 700 BC (Çakırlar and Ikram 2016). Other two types of elephants are living in African subcontinent. One is African forest elephant (*Loxodonta cyclotis*) and another one is African bush elephant (*Loxodonta africana*).

According to ENVIS Centre on Wildlife and Protected Areas, total population of Indian elephant estimation is 27,785–31,368 (in 2017). They are the largest living animal and endangered red listed species announced by IUCN from 1986. Their populations started declining as one half over the last 60–70 years i.e. three

M. Mandal and N. Das Chatterjee, *Geo-Spatial Analysis of Forest Landscape for Wildlife Management*, GIScience and Geo-environmental Modelling, https://doi.org/10.1007/978-3-031-33606-5_2

generation of their lifespan in India. The major threats are poaching, habitat degradation, habitat loss and fragmentation (Choudhury et al. 2008; Williams et al. 2020; Mandal et al. 2021a). Indian elephants restricted as major four regional groups argued by elephant field ecologist Sukumar (1992). These regions are North East India, total area covered 41,000 km^2 with 33% elephant population, East region covered 23,500 km^2 with 10% elephant population, North region area covered 5500 km^2 with 4% elephant population and South region area covered 40,000 km^2 with 43% elephant population.

In west Bengal elephants are restricted in two different regions with distinct nature of habitat. Terai region of North Bengal is considered as the natural habitat for Indian elephants but in south Bengal most of the elephants are migrated from nearest State of Jharkhand and Orissa stated by several scholars Sukumar (2003), Singh (2006), Santra et al. (2008), Kulandaivel (2010), Chatterjee (2016) and Mandal et al. (2021b). In North Bengal 488 elephants were counted whereas, in southern West Bengal, 194 elephants were counted, and the total area of proper habitat could not be estimated as reported by State Annual Report (2019–2020).

2.1.1 Elephant Habits and Their Habitat

The largest herbivore animal needs an extended foraging ground for their biological needs. This space is called 'home range'. Wild animal moves for food, keep distance from disturbance and predators, safe from difficult environmental condition and mate (Strauss 1991). Following the same pattern, elephants extend their home range as venture. From previous literature, it is seen that they move throughout their home range depending on availability of water and seasonal variation of fodder. So, ranging pattern of elephant is directed by their basic requirements such as water and food quantity and quality, their social relation and inter family competition (Sukumar 2003; Choudhury 1980). These ecological pattern and style directly related to landscape nature and characters (Desai and Hedges 2010). For this reason, the size of home range of elephants in both continent Africa and Asia differs spatially and temporally. In Tsavo West, Kenya, Africa female elephants home range size is 409 km^2, in Tsavo East 2380 km^2 reported by Leuthold (1977), Etosha-Kaokoland, Namibia 5860 km^2 stated by Thouless (1996) where in Asia, Buxa-Jaldapara Reserve, the size is 450 km^2 according to Sukumar (2003), Ruhuna National Park in Sri Lanka 100 km^2 recorded by Fernando et al. (2000). South West Bengal is the part of such extended home range of Dalma Wildlife Sanctuary, Jharkhand State and central Indian elephants (Datye and Bhagwat 1995; Chatterjee and Chatterjee 2014; Mandal et al. 2020a). Elephants naturally prefer more virgin habitats but they can also be found in small isolated populations within a highly fragmented landscape (Chatterjee 2016) in south Bengal. They can adjust themselves within a diversified landscape. Throughout India their habitat extends from mountains to plains through plateau areas. Their home range area varies from 105–155 km^2 to 650 km^2 based on the availability of fodder, water and shelter. A herd of 100 elephants would require a minimum of area of about 650 km^2 (Sukumar 2003; Chatterjee 2016).

Mega herbivores have great impact on their habitat reciprocally habitat also controlled their living pattern. Their metabolism, body size, biological behavior and social interaction vary country to country and region to region. Each elephants in Savanna region in Africa need forage of 10% of their body weight per day estimated by Sukumar (2003). Elephants in Amboseli, spend maximum time (70–75% time) for their feeding at 24 h period observed by Lindsay (1994). Normally their feeding hour started from early morning and continues to late afternoon (Baskaran 2013). Asian elephants utilize 12 h for their feeding and normal rate is of

0.8 trunkful/min (Sukumar 2003). Field expert Sukumar also stated the feeding behavioral statistics. In Asia when elephants feed on agricultural product like paddy or millets they take this require amount in just 6–7 h feeding. Elephants choose their food items by availability, taste, smell and its nutrient content. The cultivated grass or secondary vegetation such as paddy holds higher nutrient. So elephant always prefers paddy products for their daily food consumption (Plate 2.1). The mature paddy offers higher amount of nutrient i.e. 5–10% where other associated wild grasses only 2–3.8% (Sukumar 2003). Water is a very common and essential element for all living being. One Bull taker takes 100 L water at a time and 250 L per day (Sukumar 2003). In the Savannas of Africa, elephant herd migrate mile to more miles only for water during dry season. So their migration route is well demarcated from source to sink region. In fragmented landscape of Asia, elephants drink water without long distance movement due to their smaller home range and its frequent availability. But it influences their frequent movement from inside habitat to outside. Elephants of south West Bengal cannot migrate long distance to take water (Mandal 2018). In south Bengal, elephant herd prefers to stay in those forest patches where water source is available and easily accessible. Since habitat selection depends on elephant's body metabolism and hiding themselves from disturbance so they usually take rest in those forest patches of the habitat where temperature is low throughout day time in dry season. Therefore, plant composition and structure are very common factors for elephant resting and hiding.

Plate 2.1 Elephant habitat patterns near forest edge in Salboni and Lalgarh forest in Paschim Medinipur district. Photos by Rakesh Singha Dev

2.1.2 Elephant Population and Seasons of Migration

Long ranging migration of elephant is not seen in West Bengal in contrast to African elephant movement. Rather short distance movements with frequent resting are normally observed by the elephants in West Bengal. The basic habitat requirements of elephants are depending on the factors of food and water availability around their home range. In the districts of north Bengal, large strips of evergreen tropical forests with preferable number of rivers and creeks flowing within these forests become an ideal place of living for the elephants. But in south Bengal the forests are fragmented in nature hence the elephants have to cover more distance for forage.

It is evident that the migration of elephants from Dalma WLS and Mayurbhanj district in Orissa to southern West Bengal can be related to the massive land cover change in the recent past. The recent land cover changes that are beyond repair have become less favorable for the existing elephants in those areas. Availability of quality forest cover in such landscape is a strategic requirement for elephants to ignore human intervention in the western districts the state. The forest divisions of Paschim Medinipur and Jhargram districts adjacent to Dalma and Mayurbhanj are more prone to conflicts by migratory herd compared to the other districts in south Bengal. This is primarily due to better juxtaposition of the large forest patches. Another cause is proximate crop fields with the forest habitat. The districts of Paschim Midnapore, Jhargram and Bankura are equally affected by the movements of residential males from 1980s.

Another trigger factor for migration is climatic diversity. The climatic conditions of the districts of north Bengal is more stable than south Bengal. South western districts of West Bengal consists of dry deciduous forest with periods of frequent drought. The elephants are forced to migrate in search of food and water when seasonal shortage of food occurs due to consecutive dry seasons. The elephants of the south Bengal started to migrate from Dalma WLS in Jharkhand to the districts namely Jhargram, Purulia, Paschim Midnapore and Bankura with regular habit of migration. At the time of migration, elephants also start raiding the crop especially during the period of crop harvesting (Plate 2.2). They also frequently enter into the settled areas and damage assets in search of food.

Elephants also require huge supply of energy to migrate long distances. Many times infant and defeated elephants were unable to cover this long distance for return back to Dalma Wildlife Sanctuary and they used to stay in the fragmented forests. So, these groups later become the residential group of elephants in the districts of the South Bengal.

Plate 2.2 Elephants migration from Jhargram to Paschim Medinipur district using paddy fields as a habitat gap between nearest forest patches. Photo by Rakesh Singha Dev

2.1.3 Elephants in North Bengal

The Asian elephants were once roamed across the Indian sub-continent. But they are now confined to certain regions. North West Bengal comprises of the five districts of Darjeeling, Jalpaiguri, Alipurduar, Kalimpong and Cooch-Behar. This area is marked by Sikkim and Bhutan on the north, Nepal on the east, Assam on the east and Bangladesh on the south. The districts have forest cover of 3051 km^2 out of 24% of the total geographical area. The elephant home range areas of North Bengal is comprising of 1954 km^2 of forest and located to an altitude between 60 and 900 m above mean sea level. These elephant domain areas further divided into three distinct zones.

Terai: This zone is bounded by the river Teesta in the west and partly adjoins with Nepal. It comprises the forest of Kurseong Division and the Mahananda Wildlife Sanctuary.

Western Dooars: This zone is bounded by the river Teesta in the west and the river Torsa in the east. It partly adjoins with Bhutan and covers by National Pars and WLS. These are Gaurumara National Park and Chapramari Wildlife Sanctuary in Jalpaiguri, Neora Valley National Park in Kalimpong, Cooch-Behar and Jaldapara Wildlife Sanctuary.

Eastern Dooars: This zone is bounded by the river Toorsa on the west and the river Sankos on the east. This zone adjoins Assam and Bhutan. It comprises of the Buxa Tiger Reserve and Cooch-Behar Range.

In the Gazetteer of Darjeeling (O'Malley 1907) the early records of the distribution of the elephant in North Bengal were found. O'Malley stated that elephant existed in the large forested tracts at the foothills up to 10,000-ft. with three herds. These herds roamed between the Tondu forests of Jalpaiguri and a tributary of the Jaldhaka River. According to Fawcus Committee arranged by State Government stated that the elephant herd become smaller and divided in several sub groups due to tea plantation in this region. They also reported that the elephant herds sometime moved towards north into the valley of Bhutan. Their route also turned towards the east in Assam. Elephant appearance in Darjeeling district was recorded in summer. They touched this district at the time of movement on the way of Bhutan. The first systematic investigation was taken by D.K.L Chaudhury in 1977. He estimated and observed their distribution and movement pattern in North Bengal. The seasonal movement pattern and direction was pointed out by this study.

2.1.4 Elephants in South Bengal

Once upon a time, south western districts of the state were full of wildlife. From late ninetieth century to the beginning of twentieth century wild elephant herd has not appeared in State of south Bengal but a few may uncovered accidentally reported by O'Malley (1908). So disappearance of wilderness from South West Bengal was the common incident due to repeated forest fires activities extension of south eastern railway and massive tribal hunting activities (Rao 1960). Reduction of forest habitat from 1960s restricted the appearance of elephants in the western border of South West Bengal (Singh 2006). The experimental success of Arabari project by A. K Banerji, District Forest Officer (DFO) in 1971–1972 in Midnapur district, the amount and quality of the forest become enhanced. Social forestry project was set up in 1981 in West Bengal with specific objectives (Malhotra and Poffenberger 1989). After 1980 degraded low quality forest become protected and forest area also increased by social forestry programme. Mainly the Sal forest were rejuvenated in different succession (Plate 2.3). Amount of forest cover increased successively in the past three decades (Sudhakar and Raha 1994). As a result a small population of elephants of Dalma WLS from Jharkhand State ventured eastward into the State of West Bengal in 1987. Elephants were coming through the small patches of regenerated forest (Sukumar 2003).

Plate 2.3 Characterizations of tropical dry deciduous forest dominant by Sal (*Shorea robusta*) trees, South West Bengal. **a** Sapling with doted young sal, **b** dense undergrowth with young sal, **c** matured sal without undergrowth and **d** very dense undergrowth with doted matured sal tree. Sampling survey was continued from Joypur in Bankura district and Arabari forest in Paschim Medinipur district

Accidental movement of elephants from core i.e. Dalma WLS to periphery has observed repeatedly in the consecutive years. Improved ecological conditions of degraded forests attract elephants to the extended home range. Therefore, this region has become a favourable resting habitat (Plate 2.4) for elephant. The elephant newly habited in the adjoining districts like Paschim Medinipur, Paschim Bardhman, Bankura, Jhargram and part of Purulia district in southern part of West Bengal is regarded as extended home range of Dalma WLS of Jharkhand (Chatterjee 2016; Mandal et al. 2021c). The frequent movement of elephant and their colonization contemplated Government to manage them. As a result Mayurjharna Elephant Reserves with an area of 414 km^2 has been announced in 2002. This reserve is consisting of the districts of Bakura, Puruliya and Jhargram. Now there are also a few resident elephants in the region (Chowdhury et al. 1997; Mandal et al. 2020b). It is interesting to note that the recent movement of elephants in following years expanded deeper to the east (Mandal et al. 2019, 2021d).

Plate 2.4 Resting nature of a residential tusker in the edge between forest and agricultural land in Salboni forest under Paschim Medinipur district. Photo by Rakesh Singha Dev

2.2 Royal Bengal Tiger of the Sundarbans

The Royal Bengal Tiger (*Panthera tigris tigris*) is a subpopulation that ranks among the biggest wild cat alive today (Mazak 1981). This mammal has the symbolic identity in social custom and also has a glorious status in rituals background. Though, they still are listed as endangered species of IUCN Red List. It is a pride for us for considering it as the world's charismatic mega fauna (Sankhala 1978). Historically, the Royal Bengal Tiger habitat range were up to the valley of Indus River in west till nineteenth century which covered India, Pakistan, southern Nepal, Bangladesh, Bhutan and southwestern China. Heavy diversity become of their habitat squeezed and remains only in some parts the continent. Today this wild cat inhibits India, Bangladesh, Nepal, Bhutan and southwestern China (Luo et al. 2004) with confined territory. This beautiful mammal was threatened by socio-physical activities like poaching, habitat loss and fragmentation of habitat. None yet, India is the larger home where 80 percent of the tigers alive in respect of the globe. The census of 2018 claimed 2967 is the estimate number of tiger in India. Another related statistics reported that tiger population has become richer from past decade in India.

Buxa Tiger Reserve in North Bengal
West Bengal, there are two habitats of the Royal Bengal Tiger, one is the swampy mangrove forests of West Bengal and the other one is in the North Bengal in the Buxa Tiger Reserve (BTR). In the BTR region the last Royal Bengal Tiger was recorded in 1998 and 23 years later the presence was noticed again (West Bengal Forest Department 2019–2020). The ecosystem restoration activities in the BTR, including grassland management, helped bring the tiger back to the protected area.

Box 2.1 Another Tiger Reserve in West Bengal Bengal the tiger population can be found in mangrove forests of Sundarbans Biosphere and Buxa Tiger Reserve of north Bengal. 68 tigers are found as per census of 2018 in an area under 1330.10 km^2 of Sundarbans. The largest concentrated population number of the Royal Bengal Tiger experienced within this protected land which split between the two countries of India and Bangladesh. Becoming a keystone species Royal Bengal Tiger plays a critical performance in ecosystem of mangrove forests of Sundarbans. It is a tidal wetland forest delta with diversified floral and faunal composition covering an area of about 10,200 km^2 across India and Bangladesh. It also the largest mangrove forests of the world (Dasgupta et al. 2017). The mean elevation for most of the Sundarbans is less about one meter above sea level (Loucks et al. 2010). The Bengal tigers that live in this park are one of the only tiger populations adapted to living in the mangrove systems (Dasgupta et al. 2017). Varieties of living patterns of mangrove tiger separated them from main land of Asia. In 2014 National Tiger Census repotted 76 tigers were struggling in the Indian part of the Sundarbans. And become an increasing trend of their number found in 2020–21 i.e. 96 as per report of State Forest Department. The density of tigers is 12 per 100 km^2 in this area which is the 2nd highest after Jim Corbett National park which holds 14 per 100 km^2 in India. The Royal Bengal Tiger has a unique characteristic of swimming in the saline water (Plate 2.5) so their diet sometimes also include fish of various kinds. Tigers can live in a variety of environments. One of the other ideal habitats for tigers is the swamps of mangrove forest. The mangrove forests are mostly formed in the tidal waters. They create a diverse form of ecosystem that is rich in wildlife.

The Sundarbans had always been an ideal habitat of the tigers. The Sundarbans are declared as the Biodiversity hotspot as well. The Zoological Survey of India first time in 2017 took initiative to survey the number of species of flora and fauna found here. They reported Sundarbans has a rich diversity zone of its floral and faunal composition as well as micro ecosystems (Plate 2.6a, b, c and d). Such diversity of ecosystem makes an ideal place for the Royal Bengal Tiger from ancient time. The tigers prefer water during the hot summer days to cool off. So, they are located several times for their resting beside the small water puddles and ponds within the forest. Such efficient proximity to water has changed tiger's diet of Sundarbans. Sometimes they have been seen to catch and eat fish in the estuaries.

The National Park of Sundarbans was fashioned for protecting the Royal Bengal Tiger. In assurance the core areas of the mangrove forest are well demarcated and free from all kinds of human intervention. As per strict rules and regulation humans can only collect wood, honey and

Plate 2.5 Tiger in Sundarbans swimming and crossing the tidal river, a natural phenomenon. Photo by Chinmoy Barman

Plate 2.6 Different micro ecosystem zones with dominant trees under Sundarbans Biosphere Reserve

other forests products from the buffer zone of the Sundarbans National Park. Sundarbans is also famous for its indigenous fish community those have a wider value in worldwide. In spite of fishing activities are totally restricted in the core zone of the forest only to maintain a stable ecosystem. The Forest Department does regular patrolling using the launches and motorboats in order to keep the illegal poaching and theft out. None yet, Sundarbans tackles number of natural and anthropogenic challenges.

2.3 Spotted Deer of South Bengal

Another good-looking wild species is spotted dear (*Cervus axis*) also known as Chital deer or axis deer. The spotted dear is the 3rd largest deer species of India (Prater 1971) and Least Concern species listed in IUCN Red List. This species is found almost everywhere in Indian forest. They are spotted from the foothills of the Himalayas to the Peninsular India apart from the mountainous terrain. It's very concern thinking that the numbers of spotted dears have regretted immensely throughout the continent in their habitat ranges. According to National Wildlife Database Chital are now abundant only within 123 Protected Forest areas. Chital deer are sensitive social animals and they cannot be found in solitary. Female Chital is less in weight and size than the male. These males are about 30–35 in. in size with weight of 30–75 kg. Males have distingue by their horns and females without it (Schmidly and Bradley 2016). The deer group may consist of 10–150 individuals though it's depending on various situations (De and Spillit 1966; Eisenberg and Lockhart 1972; Fuchs 1997; Krishnan 1972; Schaller 1967).

They are purely herbivores so they staying in near the green foliage and thus they require huge grazing grasslands. They are well known to feed grasses, forbs and leaves of woody plants. They roam for searching long green grasses during monsoon and post monsoon time and in winter they feed fallen leaves, flowers and fruits in the Jungles. Spotted deer usually drinks water once a day but its nature more frequently in summer

(Sankar and Goyal 2004). They like to rest in the summer under the shade of trees. Their Activities become slow down during late afternoon and continues till midnight (Sankar 1994). They generally follow the same movement path for their biological requirements and do not wander in other location but accidentally it happens. They move in herds as they are most at the risk of being prey. It is very common to see that they avoid their enemy and run away from there. Predators always want to prey them and keep them the fastest choice of their diet list.

Chital deer has an exigent survival capacity so they can survive in a wide variety of habitat. Thus in Jungles of West Bengal they are commonly found. They accommodate themselves from dense forests to deciduous forests under the state. They feed mostly on the dry deciduous plants and associate undergrowth vegetation in the forest of south western districts such Jhargram, Nadia, Pschim Medinipur, Bankura, Birbhum, Burdwan and Puruliya. The district of North 24 parganas especially in Sundarbans holds mangrove it usually evergreen in nature where they have access to more species of plants (Plate 2.7).

The most reputed deer conservation sites in south Bengal are the Bethuadahari WLS in Nadia and Ballabhpur WLS in Birbhum district. Nadia district covers 3927 km^2 out of which only 12.33 km^2 is under forested area and Birbhum district covers 4545 km^2 and its recorded forest area is only 159.87 km^2. Here the forest type is tropical dry and moist deciduous forest. According to 1995 census, it had 297 spotted deer in the Sanctuary of Bethuadahari as per Annual Report Forest Department 2010–2011. Another two sanctuaries are Ramnabagan WLS in Bardhaman district and Bibhutibhushan WLS north 24-Parganas district in south Bengal. There are few population of spotted deer have remain with notified boundary.

Spotted deer are often characterized as a fickle animal in their habitat. None yet they lose their live by hunters for its soft meat and charming skin. Thus the population of these species becomes declined in incredible way. In south Bengal districts like Birbhum, Paschim Medinipur, Bankura some cultural, ritualistic or religious events conducted by tribal communities which is celebrated as "Shikar Utsav" a hunting festival. Traditionally this repeated hunting action threatens the population of spotted deer in the districts. Though the process is under control by massive awareness in spite of there is also illegal poaching by some mafia group are linked with decrease of deer population. The deer populations are not under direct threat in Nadia and North 24-Parganas due to constructive approach of society and concern authority. Since in this district deer have huge grazing ground and abundance of forest grassland that reason they can roam about freely and safely.

2.4 Crocodile of the Sundarbans

The saltwater crocodile (*Crocodylus porosus*) is the largest living reptile and have been listed as Least Concern species in IUCN Red List

Plate 2.7 Spotted Deer habitat patterns and food habits in Sundarbans

since 1996. The Sundarbans is a natural wildlife habitat including crocodile covers a range of 4260 km^2. Several canals and rivers flows inside the mangrove forest where crocodile can alive easily. A survey done by wildlife experts has estimated a population of at least 100 wild Crocodile in Sundarbans. The crocodile that found in the estuaries of Sundarbans are the saltwater crocodile. They are about maximum 6 m in length and weigh around 1000–1300 kg (Wood 1983). The saltwater crocodile is an aggressive large carnivorous.

Coastal brackish mangrove swamps and river delta areas inhabited by the crocodile of Sundarbans. Crocodile movement pattern influenced with seasonal condition especially it depends on temperature. They spend wet season in freshwater swamps and beside rivers in this area and moves towards downstream into the estuaries during the dry season. They can survive for prolonged periods in only warm temperatures and they take shelter with limited food in the warmer areas (Ross and Garnett 1989) when the cold spell hits the country. Its population can also be counted in Bhitarkanika WLS in Orissa State (Singh 2006). They are also inhabitant in parts of Andaman and Nicobar Islands (Whitaker 2008).

The Sundarbans delta that includes mangrove scrub, saltwater mixed forest and swamp forests, these patterns of landscape are ideal habitat for crocodiles (Messel and Vorlicek 1989). Crocodile becomes an aggressive predator and prey includes estuarine fish, and mammals that include deer, aquatic birds, aquatic reptiles and small species of monkeys in Sundarbans areas. They like to rest in both sand flats and mudflats (Plate 2.8). At the time of predating they hide in place of open grassland on sandy soil areas where cattle and deer are feeding. A wide variety of aquatic fauna like turtles, sea snake, monitor lizard, rat snake, prawn shrimp number of species of amphibians, fish, crabs are found in the estuaries and saltwater creeks which makes an excellent feeding grounds for the juvenile crocodiles (Ross and Garnett 1989). Thus Sundarbans Deltas region is an excellent habitat for the Saltwater Crocodile.

Year wise repeated cyclones hitting the Sundarbans have damaged more than 40% of the mangrove forest areas until. That's why many species become threatened due to only forest loss. This process forced wildlife exposes as well as crocodile from wild place to human habitation. It is very natural behave of saltwater crocodile that they have a tendency to attack human as soon as possible. Human population frequently venture into the mangrove forest inside crocodile habitat to collect firewood, honey, fish then conflict arise. Attacks on human are generally to be territorial rather than predatory in nature. In the Sundarbans, the number of fatal attacks on human has reportedly increased due to Habitat loss and degradation. Unfortunately, Forestry Department not actively addressing the problem of crocodile attacks in the Sundarbans because of its low intensity than human-tiger conflicts. But it keeps in mind crocodiles are both dangerous in land as well as in the water. Both the primary

Plate 2.8 Saltwater crocodile's habitat in Sundarbans

(fatal) as well as secondary (non-fatal) attacks should be taken into consideration immediately to address the issue of man-crocodile conflict.

2.5 Wild Boar of South Bengal

The Indian boar (*Sus scrofa cristatus*) has a typical shape of a domestic pig with a long, blunt snout, small eyes, and large ears (Amills et al. 2017). They are more long-legged and appear more powerful and violent than domestic pigs. They may also famous for their small hump on the shoulder. They are usually brown, but can appear rusty-red or black in different habitat. Indian boars have tusks both males and females although these are almost longer in males. The wild boar is adapting them in an environment with distinct habitat and consumes nearly any type of food. They are omnivores in nature. They also are an excellent swimmer found in Sundarbans (Plate 2.9) and a very fast runner. At the time of grazing a peculiar sound releases to know their presence. They are very active during the night time. This animal spends as much as 12 h per day for sleeping in a nest. Females of this species exhibit a social behaviour, loosely organized groups of 6–30 individuals. It varies from region to another region. Two or more groups may occasionally share the same area without mixing each other and competition. Male wild boars tend to lead solitary life during the most of the year. They socialize only in the reproductive season.

As the population of the wild boars is not endangered by IUCN and as they can alive in almost all condition. According to the forest report of Kerala, around 58,000 wild boars are counted there at present in the state (Report 2022). However, in the State of West Bengal no such census has been carried out that can estimate its population.

Another fact associate with this animal, they are known as the farmers of the forest. This is because they dig up soil in search of foods under soil surface and as a result they till the forest soil. Wild boars cause significant damage in agricultural land especially in the areas where the tribal communities live with their farms in south Bengal. This community inhabits in places in a close proximity to the forests in the districts of Jhargram, Bankura, Paschim Medinipur and Purulia. They have very less amount of land and practice subsistence nature of farming. The boars frequently enter in to these agricultural lands and destroy the vegetations. Tribal communities of south-western West Bengal hunt wild animals almost all the year round. However, certain days that generally coincide with some cultural, ritualistic or religious events, are celebrated as "hunting festivals" (HEAL Report 2020). Ritualistic hunting is possibly one of the biggest killers of wildlife in South-West Bengal as well

Plate 2.9 Wild boar habitat pattern in Sundarbans

as wild boar. The problem is most acute in the southern districts of Purulia, Jhargram, Bankura, and West Midnapur.

2.6 Leopard in North Bengal

The Indian leopard (*Panthera pardus fusca*) is listed in Vulnerable species in IUCN Red List. This big cat species is found widely in Indian sub continent. Estimated 7910 leopard counted and they live in around tiger reserve in India except Northeast. It is speculated that an around 12,000–14,000 leopard alive in the Indian sub continent. State of Madhya Pradesh holds the largest number i.e. 3421 as per 2015 report. Like other discussed wildlife (Fig. 2.1) leopard's presence in West Bengal also become abundant with 164 in 2004 reported by State Annual Report 2019–2020 in northern districts after that leopard census is not carried out until. Leopard in south Bengal become regionally extinct. Yet, the

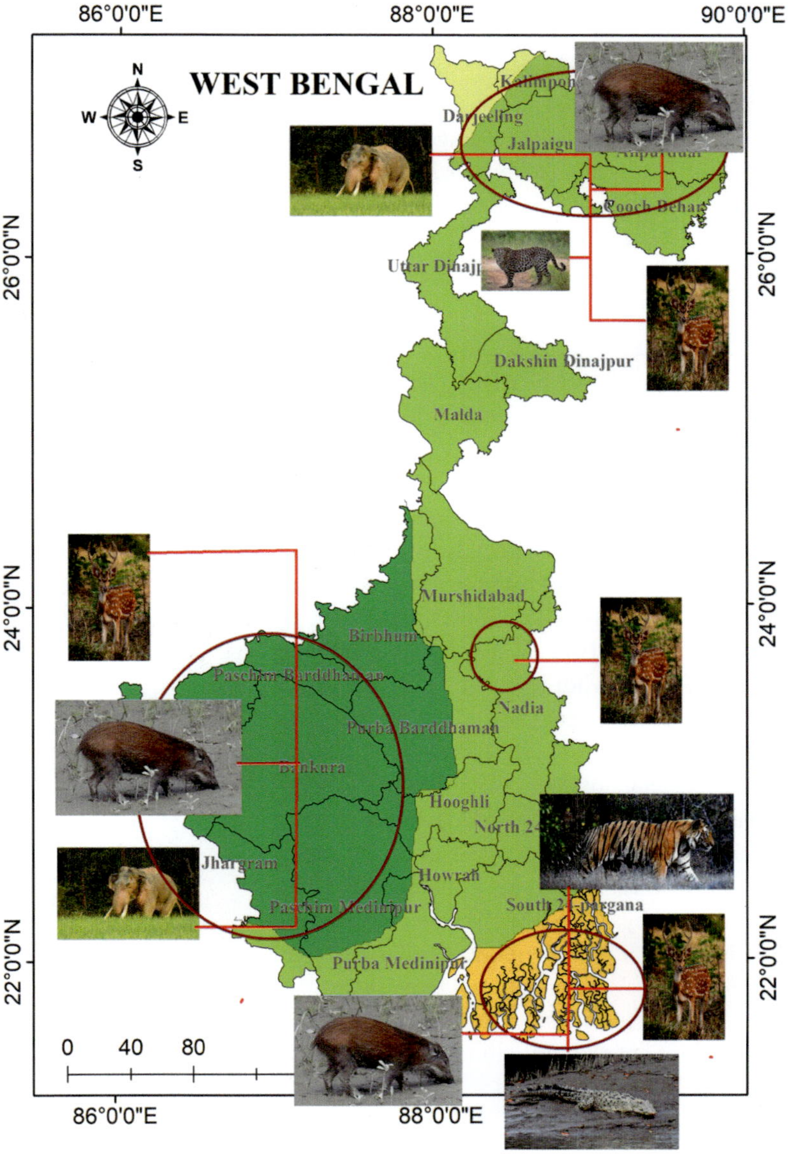

Fig. 2.1 Zone wise schematic presentation of wild animals in West Bengal

district Gazetteer Report by O'Melley (1908) said about their existence in the western part of Raipur Forest Range under Bankura district at the beginning of twentieth century.

Human death and property loss leopard attack become regular in north Bengal (Kshettry et al. 2017; Borthakur et al. 2021). They had previously reported about attacks on human being by leopards generally happening during day time in tea estate garden areas in the Jalpaiguri district of North Bengal. Other scholars Naha et al. (2018, 2020) predicted risk on humans in the central and the western parts of North Bengal by leopard attack. Further they also reported livestock predation is higher in between winter and spring season in the day time in North Bengal. The human-leopard conflict in North Bengal is often characterized to massive change in land use alteration. Generally the expansion of tea gardens in the region are the major cause of it (Manoj et al. 2013; Vyas and Sengupta 2014; Naha et al. 2018).

2.7 Unexpected Ecological Migration of Wildlife in South Bengal

Recently rare ecological events occur in human habited areas of south Bengal. Once upon a time the culturally modified landscape were comes under the boundary of Janglemahal as stated by O'Mealy (1908, 1911). At that time sporadic appearance of wildlife were not surprising in between the areas of Janglemahal and Bhagirathi-Hooghly River. Presence of wildlife in the ancient folklore like 'Sitala Mangal', 'Manasa Mangal', 'Sasthi Mangal' prove their appearance in these areas. And even older persons in this region when remembrance their childhood, they said several kinds of wildlife experience which were more terrible to them. However these enormous folklores lost their glory. But it is surprising to see that in recent past some migration of wild animal become crucial event to the ecologist, forest officials as well as forest fringe dwellers in south Bengal. Such unexpected ecological events are given in the next section in this chapter.

Box 2.2 Unexpected Longest Ecological Migration of Royal Bengal Tiger in South West Bengal in Recent Time

January 30, 2018: On this day, the first tiger was heard to be found in Jangalmahal. An unknown animal attacked some cows grazing in the forest in Madhupur area of Lalgarh. Cows have deep claw marks on their bodies and necks which was very uncommon for forest dwellers. From this incident, the presence of a tiger in the forest spread among the villagers. Although at that time the people of the area suspected this incident to be a wolf or hyena attack because they have not any experience for appearance of tiger in this forest.

February 14, 2018: After that, there were reports of seeing the footprints of an unknown animal from the area around Madhupur forest fringe. Many people say and claim these are tiger footprints. After that, pictures and news of tiger footprints in the forest fringe bring front of us with wildfire on social media. The people of the area were so much terrified that many schools near the jungles of Lalgarh were partially closed after this incident. Local people stopped going to the forest for basic needs.

February 27, 2018: The Forest Department has entered the field to deal with the situation for controlling that. Talking to eyewitnesses from dwellers, a total of seven trap cameras were installed at different places in Melkheria forest near Lalgarh and Madhupur forest. Meanwhile, a herd of elephants in the forest created problems by breaking some trap cameras.

Picture source West Bengal Forest Department

March 2, 2018: Finally, a photograph of a full-grown male Royal Bengal Tiger was found in the forest department's leaf camera trap in Melkheria forest of Lalgarh. D. F. O. of Midnipur Forest Department Rabindranath Saha held a press conference and disclosed the truth of this incident of tiger. He said that there is no doubt that the animal seen in the picture is a tiger. The disclosures and telecasting of the presence of tiger in Jungalmahal become big news that surprises from expert level to local dwellers level. It is initially assumed that the tiger may have migrated here from Simlipal in Odisha. The images obtained from the trap camera are sent to the National Tiger Conservation Authority (NTCA) for testing to identify the tiger. The forest department conducts miking in the villages near the forest and distributes leaflets prohibiting the local people from entering the forest.

March 3, 2018: Special cages and teams of experts arrive from Sundarbans to capture tigers. Goat-baited cages are littered around the forest where the tiger picture was found. Although the tiger was not caught in the cage, tiger droppings were found in front of the cage in Madhupur forest. A subsequent examination of the droppings revealed the presence of wild boar hair. This means that the tiger is hunting and eating wild boar in the forest. So instead of goats, pigs are given in cages as bait.

March 5, 2018: The National Tiger Conservation Authority (NTCA) reports that the skin spot pattern of a tiger captured by a trap camera in Lalgarh did not match that of a tiger captured in Simlipal in the recent past. Along with Simlipal, the name of Palamu in Jharkhand came up in the discussion of experts as the possible habitat of the tiger coming to the forest of Lalgarh. Three possible routes for the tiger to come to the Lalgarh forest came up in the opinion of experts. Odisha Simlipal to Goru Mohishani to Dumaria via Subarnarekha to across Kansavati. The tiger may have entered Lalgarh through Jhargram forest. The tiger may have come to Lalgarh from Palamu in Jharkhand via East Singhbhum to Hatibari. From Jharkhand's Dalma to Belpahari's Kangra Jhor to reach Lalgarh's forest.

13th April 2018: The incident which started with reports of sighting of a tiger by villagers in Lalgarh forest in January 2018, resulted in the recovery of many arrows pierced Royal Bengal Tiger in the forest of Bagghara in West Mednipur on 13th April 2018. The poor stray tribe tiger was killed and proved that this forest was his hideout. Not being able to find the tiger physically for more than two months and finally its tragic end is considered to be the biggest advertisement of the failure and inaction of the West Bengal Wildlife Department near to the world by all concerned including wildlife conservation experts. Social media has become full of protests and protests. The hearts of Jangalmahal are saddened by this unfortunate unexpected incident.

Various questions have arisen around the death of Royal Bengal in Lalgarh. As everyone knows, the tiger was not completely unknown from few days. At times he has made his presence known in various places. Sometimes he came back after seeing a caged goat, sometimes he took refuge in a canal pipe and was almost trapped. In the last one-and-a-half months, drones circled the sky during the Baghbandi operation, and two forest workers died after staying awake all night. However, he was not captured alive. Why the forest department could not catch the tiger. Why can't they avoid the tiger-man conflict? After the incident, a section of forest personnel admitted that since tigers do not come into the forest in this area, they do not have the training to chase or cage them. Environmentalists and wildlife experts have informed the local tribal that the incident in the name of hunting is very harmful for the environment and conservation. The forest department staff also feels that if the forest was calm, the tiger would not have lost its life in the tiger forest.

Source Rakesh Singha Dev

Box 2.3 Unexpected Longest Ecological Migration of Nilgai in South West Bengal in Recent Time

Unexpected ecological migration of Nilgai

The number of Nilgai (*Boselaphus tragocamelus*) are very low in the world today. Nilgai has become an almost endangered species by IUCN. Nilgai is generally not found in the natural habitat of South Bengal. But in 26th October, 2022 it was seen in Birbhum district. The local dwellers have heard the name of Nilgai, but few have been lucky enough to see them. A Nilgai was seen wandering in a village called Avinashpur under Parui police station in Birbhum. Everyone thought that it may be a deer that had come from somewhere. Initially, it was assumed that the Nilgai had escaped from the Jharkhand region and somehow entered the Birbhum area. A team of four forest department personnel from Dubrajpur, Birbhum, Siuri and Bolpur tried for five and a half hours to rescue the Nilgai. And finally they manage to catch it. The forest department informed that the rescued Nilgai will be rehabilitated in its suitable environment soon. And vigilance will also be kept so that it does not reenter the locality.

The similar case was found in Bishnupur, Bankura district in 1st September, 2021. Villagers of Kulupukur village of Bishnupur, Bankura were surprised to see the small horse-like animal in the shrimp farming field near the forest in the early morning. After getting information from them, the forest workers came. They said that the animal was a Nilgai. After a lot of effort, they finally rescued the Nilgai. DFO (Bishnupur Panchet Division) Satyajit Roy said, "I don't know how the full-grown male Nilgai came to our forest". He said, after first aid, the Nilgai was released in the deep forest of Jaipur in Bankura and kept under observation for a few hours. But the Nilgai was coming out of the forest again. The DFO said, "The Nilgai was sent to the Alipore Zoo on the instructions of the higher authority keeping in mind the safety of the animal."

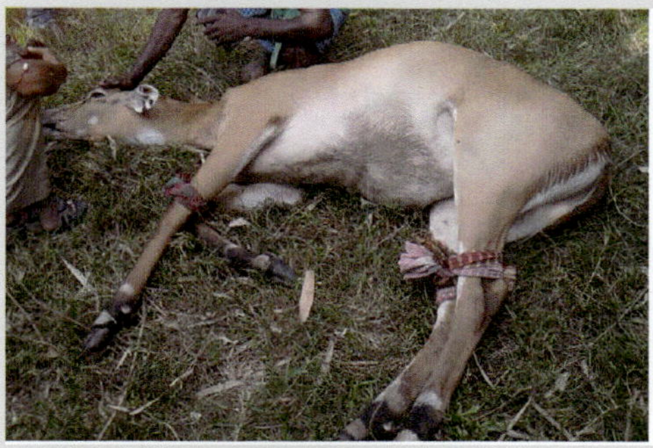

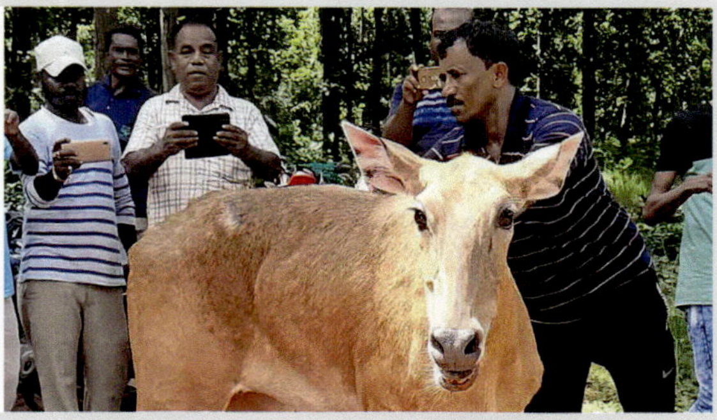

Picture source West Bengal Forest Department

Box 2.4 Unexpected Ecological Migration of Baby Crocodile

On 12th and 24th September 2020, two hatchlings of crocodiles were caught in the nets of fishermen in Khejuri estuary area of Purba Medinipur. A hatchling of almost the same age was caught in a fisherman's net in the Bagui river of Patashpur in 26th September, 2020. But how was the crocodile reached 40–50 km away from Khejuri, that was a question for local communities and environment activists. The forest department has advised locals to be cautious as there may be mother crocodiles near the hatchling. Balaram Panja, assistant officer of the district forest department, said, "A baby crocodile was found in Bagui river near Tapinda of Gokulpur village panchayat of Patashpur". Crocodiles have never been seen in this area before. He said, "It is not clear how the baby crocodile came here. It seems that a crocodile had laid eggs in the coastal area during the Covid 19 period. Afterword those eggs hatch, they disperse into local rivers and canals in search of food. As the reason, he said, "The parents do not help the crocodiles to grow up. After the eggs hatch, the babies have to fend for themselves".

Somnath Das Adhikari, President of Biodiversity Management Society of Potashpur 1 Block, said "we think a crocodile came here from the Sundarbans and laid eggs here.' He also said, the Keleghai river has joined the Haldi river. Again Bagui river joins Keleghai near Tulsichara in Potashpur. As a result, the baby crocodile entered Keleghai river from Khejuri estuary through Haldi river estuary. Then baby crocodile came to the Bagui river in search of food".

Box 2.5 Unexpected Ecological Migration of Hyena

A full-grown male Hyena was beaten to death by villagers after it entered the settlement area. The incident happened in Gokulpur of Potashpur police station area in Bagui river basin area in 2nd February, 2016. According to the State Biodiversity Board, the Hyena is also known as Lakra or Adhabagha. They are scattered here and there in scrubby fields, ditches or stone mounds. They like all these environments but stay away from locality. They also hide in crops or bushes. They make the fox hole bigger and rest in it. It usually hides during the day but comes out to hunt at night. In search of prey, it moves very fast to a distance of several miles. They prey on domestic animals as well as dead animals. In rural Bengal, at one time it was also called child lifter. Somnath Das Adhikari, President of Patashpur-1 BMC said, "This animal is not seen in the ecosystem here. Supposedly, they came to this area from the forest area of West Medinipur in search of food". East Medinipur district forest officer Shyamal Chakraborty said, "They are protected under the third level of importance in the Indian Wildlife Protection Act. Harm them is a punishable offence. The entire incident will be investigated and legal action will be taken." According to local sources, this animal was seen in the surrounding villages since last fifteen days from occurrence. Madhusudan Das, a member of the local Panchayat Samiti said, "Chickens and goats were missing from the village for the past few days". Madhusudan Das was attacked by a Hyena while going to work in the field in the morning. As soon as the animal came to light, the villagers killed him." On this day, after killing him, it was hanged in the field in front of a village house. Thousands of people gathered to see it.

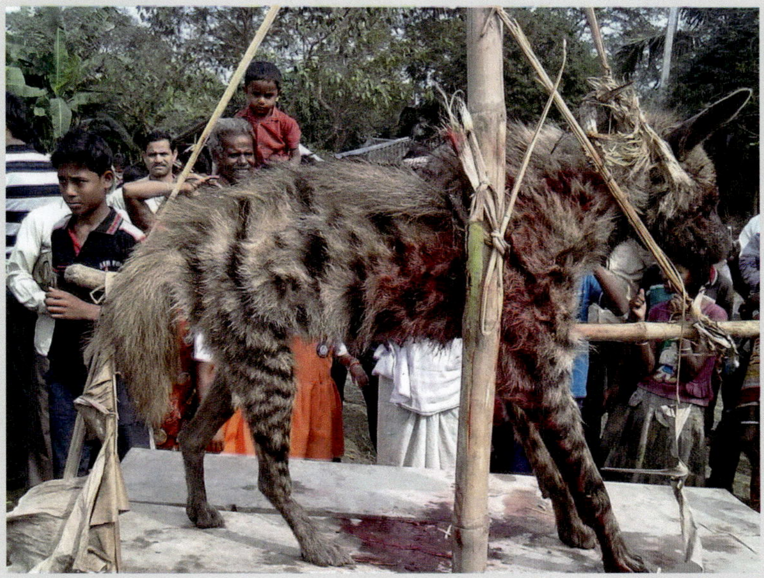

In 12 November, 2018 a railway worker Ashok Das saw a bloody Hyena walking towards the village next to the line. His face is also bleeding. The Hyena was injured by a train while crossing the railway line. The Hyena crossed the railway line towards the National Highway No. 60 and entered the village of Doharpur in Narayangarh. Suddenly, seeing the Hyena, the fear of the tiger spread in the village. The Hyena also had injuries on its mouth, tongue and teeth. A few brave youth among the villagers caught the Hyena and informed the forest department. Mintu Chakraborty, staff of Narayangarh beat office in Belda range, and others came and sent the hyena to Hijli Rescue Centre for treatment.

Akhil Singh, Dilip Singh, Samaresh Singh, Bimal Singh, Sujit Singh, Jagannath Singh went out to catch the Hyena in the morning. They chased behind and caught the Hyena near the bank of Chandla Pond. The Hyena could not run too fast because of the paddy field. Their statement, "Even though we were afraid at first, we tried to catch the Hyena after we found out that it was injured. I will inform the police later. The forest department came and took away the hyena. Panchayat member of the area Indrajit Singh said, "We did not allow the Hyena to be killed. I have handed over the live Hyena to the forest department".

D. F. O. Arup Mukhopadhyay said, "The Hyena has come out of the forest for." The Hyena was injured by the train. If it was healthy, it will be released into the forest. Otherwise it will be sent to Jhargram or Alipur zoo.

Picture source Anandabajar Patrika

Box 2.6 Unexpected Ecological Migration of Gangetic Dolphins in Bhagbanpur, Purba Medinipur

15th November, 2019: Gangetic dolphins are an endangered species blamed for human activities. Somehow it got into the canal in Bhagbanpur of Purba Medinipur which was a surprising incident for the local people. As soon as the locals saw it and informed the police and forest department. The forest workers tried to return the dolphin to the river. But, due to the fish nets in the water of the canal, its speed was repeatedly interrupted. Dolphin was going towards the river avoiding it. But, the water was so polluted that the dolphin could not be saved, forest officials said. Local forest officer Swagta Das said, "I was trying to get the dolphin back into the river. It also crossed 15 km from the main river. But, the fishing nets in the canal stuck the way. The canal water became highly polluted due to dumping of plastic. Therefore Dolphin was died, it could not be saved."

Box 2.7 Unexpected Ecological Migration of Gangetic Dolphins in Bhagbanpur, Purba Medinipur

Several issues come out form unexpected occurrence of wild animal in south Bengal. Shrinkage of natural habitat due to cultural pressure may be on of the factors for such incident but most important factor is the alteration of land characters (Mandal et al. 2021a). These characters directly determining the long ranging wildlife movement behaviour as discussed in tiger movement in south Bengal forest and elephant movement behaviour (Mandal et al. 2021c) that will be elaborately discussed in the coming chapters. Though south Bengal historically is not a natural habitat for elephants (Chatterjee 2016), it is an extended home range of elephant from Dalma WLS (Mandal et al. 2015, 2021d) but from few decades the jumbos are going towards east from continuous forest patches to fragmented forest patch. As a result frequent news's are broadcast as entitled with 'elephant entree in a new village again' or in Bengali words " দলমার হাতি দাপিয়ে বেড়াচ্ছে নতুন লোকালয়". So, all these ecological issues may be handled in an inclusive way coupling the landscape character, wild animal ecology and socio-cultural background of the human population.

References

Amills M, Megens HJ, Manunza A, Ramos-Onsins SE, Groenen MA (2017) A genomic perspective on wild boar demography and evolution. In: Ecology, conservation and management of wild pigs and peccaries, pp 376–387

Baskaran N (2013) CEPF Western Ghats special series: an overview of Asian Elephants in the Western Ghats, southern India: implications for the conservation of Western Ghats ecology. J Threatened Taxa 5(14): 4854–4870

Borthakur U, Das PK, Deb M, Maheen R, Neog A, Das AK, Barua B and Biswas A (2021) Leopard (Panthera pardus) ecology in North Bengal including population estimation, distribution, habitat use pattern and human-leopard conflicts. Aaranyak, Tech Rep 1–87

Çakırlar C, Ikram S (2016) When elephants battle, the grass suffers. Power, ivory and the Syrian elephant. Levant 48(2):167–183
Chatterjee ND, Chatterjee S (2014) Changing Habitat and Elephant Migration from Dalma Wildlife Sanctuary, Jharkhand to Panchet Forest Division, Bankura, West Bengal: a biogeographical analysis. In: Climate change and biodiversity. Springer, pp 209–222
Chatterjee ND (2016) Man-Elephant conflict: a case study from forests in West Bengal. Springer, India
Choudhury DKL (1980) An interim report on the status and distribution if elephants in north-east India. In: The status of the Asian Elephant in the Indian subcontinent. IUCN/SSC Report, BNHS, Bombay, pp 43–58
Chowdhury S, Khalid MA, Roy M, Singh AK, Singh RR (1997) Management of elephant populations in West Bengal for mitigating man-elephant conflicts. Wildlife Institute of India, Dehradun
Choudhury A, Lahiri Choudhury DK, Desai A, Duckworth JW, Easa PS, Johnsingh AJT, Fernando P, Hedges S, Gunawardena M, Kurt F, Karanth U, Lister A, Menon V, Riddle H, Rübel A, Wikramanayake E (2008) IUCN SSC Asian Elephant Specialist Group. Elephas maximus. IUCN Red List of Threatened Species. e.T7140A12828813
Dasgupta S, Sobhan I, Wheeler D (2017) The impact of climate change and aquatic salinization on mangrove species in the Bangladesh Sundarbans. Ambio 46 (6):680–694
Datye HS, Bhagwat AM (1995) Home range of elephants in fragmented habitats of central India. J Bombay Nat Hist Soc 92(1):1–10
De R, Spillit J (1966) A study of the chital or spotted deer in Corbett National Park, Uttar Pradesh. J Bombay Nat Hist Soc 63:576–598
Desai A, Hedges S (2010) Notes from the Co-chairs IUCN/SSC Asian Elephant Specialist Group. Gajah 33(2010):3–5
Eisenberg JF, Lockhart M (1972) An ecological reconnaissance of Wilpattu National Park, Ceylon
Fuchs ER (1997) Behaviour. In: Ables ED (ed) The axis deer in Texas. Caesar Kleberg, Research Program and Texas A and M University, Texas, pp 24–52
Fernando P, Pfrender ME, Encalada SE, Lande R (2000) Mitochondrial DNA variation, phylogeography and population structure of the Asian elephant. Heredity 84(3):362–372
Krishnan M (1972) An ecological survey of larger mammals of peninsular India. J Bombay Nat Hist Soc 69:469–501
Kshettry A, Vaidyanathan S, Athreya V (2017) Leopard in a tea-cup: A study of leopard habitat-use and human-leopard interactions in north-eastern India. PLoS ONE 12(5):e0177013. https://doi.org/10.1371/journal.pone.0177013
Kulandaivel S (2010) A paradigm shift in the elephant depredation in South Bengal. Divisional Forest Officer, Bankura North Division, Personal Communication
Leuthold W (1977) Spatial organization and strategy of habitat utilization of elephants in Tsavo National Park, Kenya
Lindsay WK (1994) Feeding ecology and population demography of African elephants in Amboseli. University of Cambridge, Kenya
Loucks C, Barber-Meyer S, Hossain M, Abraham A, Barlow A, Chowdhury RM (2010) Sea level rise and tigers: predicted impacts to Bangladesh's Sundarbans mangroves. Clim Change 98(1):291–298
Luo SJ, Kim JH, Johnson WE, Walt JVD, Martenson J, Yuhki N, O'Brien SJ (2004) Phylogeography and genetic ancestry of tigers (Panthera tigris). PLoS Biol 2(12):e442
Malhotra KC, Poffenberger M et al (1989) Forest regeneration through community protection. In: Proceedings of working group meeting on forest protection communities, Calcutta. Merriam G (1984) Connectivity: a fundamental ecological characteristic of landscape pattern. In: Brandt J, Agger P (eds) Proceedings of the first international seminar on methodology in landscape ecological research and planning. International association for landscape ecology, Roskilde University Centre, Roskilde, Denmark, pp 5–15
Mandal M (2018) Forest range wise Asian Elephant's (Elephas Maximus) Habitat suitability assessment through Food and Water availability: a case study in Panchet forest division, Bankura, West Bengal. Int J Basic Adv Res 2456–1372. ISSN 2454-4639
Mandal M, Chatterjee ND (2019) Forest core demarcation using geo-spatial techniques: a habitat management approach in Panchet Forest division, Bankura, West Bengal India. Asian J Geogr Res 2(2):1–8
Mandal M, Chattarjee ND (2020a) Land use alteration strategy to improve forest landscape structural quality in Radhanagar forest range under Bankura district. Eurasian J Forest Sci 8(1):1–10
Mandal M, Chatterjee ND (2020b) Elephant's habitat suitability assessment through geo spatial quantification in Panchet forest division, West Bengal. Ecofeminism Clim Chang
Mandal M, Chatterjee ND (2021a) Geospatial approach-based delineation of elephant habitat suitability zones and its consequence in Mayurjharna Elephant Reserve, India. Environ Dev Sustain 23(12):17788–17809
Mandal M, Chattarjee ND (2021b) Estimation of forest ecosystem quality using GIS tool in Panchet forest division, West Bengal, India. In: Forest resources resilience and conflicts. Elsevier, pp 203–213
Mandal M, Chatterjee ND (2021c) Forest landscape and its ecological quality: a stepwise spatiotemporal evaluation through patch-matrix model in Jhargram District, West Bengal State. India. Regional Sustainability 2(2):164–176
Mandal M, Chettarjee ND (2021d) Human-Elephant conflict in Joypur forest influence areas, West Bengal, India. Gajah 54:34–36
Mandal M, Chettarjee ND, Hazra J (2015) Elephant migration and colonization in Bankura district, West

Bengal, India. Vidyasagar University. Int J Geogr Environ 14:46–52
Manoj K, Bhattacharyya R, Padhy PK (2013) Forest and wildlife scenarios of Northern West Bengal, India: a review. Int Res J Biological Sci 2(7):70–79
Mazak V (1981) Panthera tigris. Mamm Species 152:1–8
Menon V, Tiwari SK (2019) Population status of Asian elephants Elephas maximus and key threats. International Zoo Yearbook 53(1):17–30
Messel H, Vorlicek GC (1989) Ecology of Crocodylus porosus in northern Australia. In: Crocodiles: their ecology, management and conservation, pp 164–183
Naha D, Dash SK, Chettri A et al (2020) Landscape predictors of human–leopard conflicts within multi-use areas of the Himalayan region. Sci Rep 10:11129. https://doi.org/10.1038/s41598-020-67980-w
Naha D, Sathyakumar S, Rawat GS (2018). Understanding drivers of human-leopard conflicts in the Indian Himalayan region: Spatiotemporal patterns of conflicts and perception of local communities towards conserving large carnivores. PLoS ONE 13(10):e0204528. https://doi.org/10.1371/journal.pone.0204528
O'Malley LSS (1907) Darjeeling district gazetteer. Government of West Bengal
O'Malley LSS (1908) Bengal district gazetteers: Bankura. Bengal Secretariat Book Depot, Calcutta
O'Malley LSS (1911) Bengal district gazetteers: Midnapore. Bengal Secretariat Book Depot, Calcutta
Prater SH (1971) The book of Indian Animals. Bombay Natural History Society, Mumbai
Rao V (1960) One hundred years of Indian Forestry, Vol II (Forests), Forest Research Institute, Dehradun
Ross CA, Garnett S (eds) (1989) Crocodiles and Alligators. Checkmark Books. ISBN 978-0816021741
Sankar K (1994) The ecology of three large sympatric herbivores (chital, sambar, nilgai) with special reference for reserve management in Sariska Tiger Reserve, Rajasthan. Ph. D. Thesis. University of Rajasthan, Jaipur
Sankar K, Goyal SP (2004) Ungulates of India. ENVIS Bulletin: Wildlife
Sankhala K (1978) Tiger: the story of the Indian Tiger. Collins, Glasgow
Santra AK, Pan S, Samanta AK, Das S, Halder S (2008) Nutritional status of forage plants and their use by wild elephants in South West Bengal, India. Trop Ecol 49 (2):251
Schaller GB (1967) The Deer and the Tiger. University of Chicago Press
Schmidly DJ, Bradley RD (2016) The mammals of Texas. University of Texas Press
Shoshani J, Eisenberg JF (1982) Elephas maximus. Mammalian species
Singh AK (2006) Ecological investigation of human-elephant conflicts in South West Bengal. Saurashtra University
Strauss SY (1991) Indirect effects in community ecology: their definition, study and importance. Trends Ecol Evol 6(7):206–210
Sudhakar R, Raha AK (1994) Forest change detection study of nine districts of West Bengal through digital image processing of Indian Remote Sensing Satiate data between 1988 and 1991—Procedural Manual and Inventory. Regional Remote Sensing Service Center, Kharagpur and Forest Department, Govermentt of West Bengal Joint Collaborating Project
Sukumar R (1992) The Asian elephant: ecology and management. Cambridge University Press
Sukumar R (2003) The living elephants: evolutionary ecology, behaviour, and conservation. Oxford University Press
Thouless CR (1996) Home ranges and social organization of female elephants in northern Kenya. Afr J Ecol 34 (3):284–297
Vyas P, Sengupta K (2014) Human-leopard conflict in North Bengal India. Tigerpaper 41(1):1–6
Whitaker N (2008) Survey of human/crocodile conflict in the Union territory of the Andaman Islands, Hut Bay, Little Andaman, Jan 2008. Madras Crocodile Trust, Madras
Williams C, Tiwari SK, Goswami VR, De Silva S, Kumar A, Baskaran N, Yoganand K, Menon V (2020) Elephas maximus. The IUCN red list of threatened species 2020: e. T7140A45818198
Wood G (1983) The guinness book of animal facts and feats. Sterling Pub. Co. Inc., New York

3 Relationship Between Forest Habitat Composition and Wildlife from South West Bengal

Abstract

Scientists are often interested in searching for the forces that cause the distributions of animal species among diversified environments. The responsible factors vary frequently subjects to time and space. An animal's habitat is, in the most general sense, the place where it lives. Animals prefer to live in an area where the basic resources such as food, water and shelter are present. The animals have to adapt themselves with the changing habitat due to climatic extremes, anthropogenic intervention and the competitors and predators. Humans occupy nearly all terrestrial surfaces of the earth, but other species of animals are restricted to particular places or niche or habitat. This gives rise to a typical habitat wildlife relationship over the changing landscapes. Habitat–wildlife relationships help current levels of understanding and assessing the relative importance of environmental features in the abundance of organisms. The primary purpose of this chapter is to review some of the habitats of animals in the study area and their relationship with that habitat and to outline how these components changed over time.

3.1 Nature of the Tropical Dry Deciduous Habitat

South West Bengal is the extended part of Chota Nagpur Plateau. Geomorphologically, the region slopes down towards east. Rapid undulation occurs with sudden low height peak in the western part which makes a beautiful diversified topography. Landscape Western part of South Bengal and its adjoining areas are rich in wildlife and wilderness (O'Malley 1908, 1911). Malley also stated that the region has dense forest cover during colonial period. For that reason, the surrounding areas of these districts are known as 'Jungle Mahal'. After independence, enormous forest destruction rapidly changed the existing landscape setup (Chowdhury et al. 1997). These unethical activities restructured the forest patch (Das Gupta et al. 1989; Gupta et al. 2019). Forest destruction has taken place due to several societal demands such as agricultural expansion, industrial setup, new colonies, road network development and high demand for forest products. These activities destroyed the entire ecosystem and degraded the ecological balance. But after 1980, forest demolition intensity gradually improved. Several programmes were implemented to protect and improve forest. Many

M. Mandal and N. Das Chatterjee, *Geo-Spatial Analysis of Forest Landscape for Wildlife Management*,
GIScience and Geo-environmental Modelling, https://doi.org/10.1007/978-3-031-33606-5_3

afforestation and re forestation programmes were introduced in this area to increase forest cover and improve forest quality in open land of adjacent forest areas. One of the most successful attempts was the implementation of Joint Forest Management program (Debnath et al. 2006). After the experimental success of Arabari project in 1971–1972 in Paschim Midnapur district, a social forestry program was introduced in 1981 in West Bengal with different objectives (Malhotra et al. 1989). The main two objectives were forest-based employment generation and sharing of forest product for forest dependent dwellers. These actions exhilarate the forestry program (Palit 1989, 1991). Successful launching of Joint Forest Management program in 1980, the degraded forest becomes not only protected but also forest area increased by social forestry in the barren lands of South Bengal. Thus the Amount of forest cover had been increased radically from the past three decades (Sudhakar et al. 1994). The forest qualities also increased after matured growth of the planted trees, especially sal (*Shorea robusta*) in these areas that endorse wildlife activity.

At the beginning of twenty-first century, wildlife and their activities have once again started in the forest of several districts in South Bengal (Chatterjee et al. 2014; Mandal and Chatterjee 2019). Wildlife habitation, migration, colonization, predation, etc. can be observed as usually seen in the past decades. Now the question arises why these activities are started and why wildlife has chosen this area for their residence again? Changing landscape may be the principal cause behind this phenomena as mentioned by the researchers all over the world (Fahrig and Merriam 1985; Forman 1995; Gustafson 1998; Farina 2006; Cushman 2006; Dramstad et al. 1996; Fernando et al. 2008; Kumar et al. 2008; Chatterjee 2016; Mandal and Chatterjee 2018). Habitat composition is characterised by the response of biological components the response of such as nature of colonization (MacArthur and Wilson 1967), rate of extinction (Li and Wu 2007), movement and ranging patterns (Murcia 1995) of survival species etc. These outlined responses are determined by three basic components of landscape these are structural composition, species composition and locational composition of forest habitat (Fig. 3.1). The present chapter evaluates and analyzes the forest landscape characters in this region for understanding the ecological behaviour as a whole.

3.1.1 Structural Composition

Landscape is a key unit for biodiversity as well as wildlife management. Natural land unit such as forest, wetland, barren land with wild grass community in a landscape is the most considerable elements to the ecologists as well as policymakers (Dunning et al. 1992; Harrison and Fahrig 1995). These virgin lands accomplish with corresponding cultural landuse that creates homogeneity or heterogeneity. Landscape heterogeneity is an essential subject matter in landscape ecological research (Lovett et al. 2005; Cushman et al. 2010). Generally, more inhomogeneous diversified landscape makes more complexity (Forman 1995; Arroyo Rodríguez et al. 2017). Fragmentation, regeneration and isolation processes are responsible for such kind of heterogeneity in the landscape. Homogeneous land unit is scattered into smaller unit due to fragmentation and isolation process (Merriam 1984; Kotlair and Wiens 1990; Naveh 1994; Forman 1995; Pickett and Cadenasso 1995; Mandal and Chatterjee 2021c). It is interesting to see that sometimes regeneration process like forest regeneration, wetland reclamation improves landscape homogeneity as argued by Chazdon and Guariguata (2016). Historically, anthropogenic activities isolated the virgin land into dissimilar land units (Cushman 2006). These processes break down the existing landscape structure and composition. That's why every single land unit i.e. patch plays a significent role in ecological performance of species (Farina 2008; Forman 1995; Lovell and Johnston 2009). When the corresponding patches are highly proximate to each other the supremacy is determined by ecological function and behaviour over the entire landscape. e.g. forest patches in an

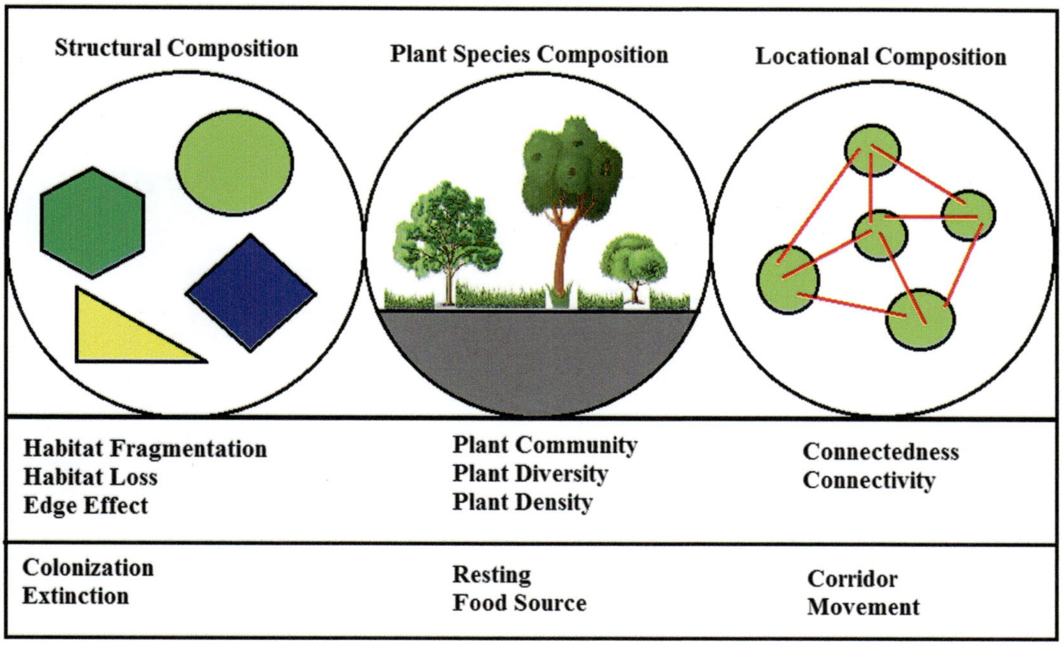

Fig. 3.1 Diagram representing landscape composition to understand ecological patterns and processes

agricultural matrix. For this reason forest patch is measured as an appropriate habitat for wildlife in every wildlife domain (Sitompul et al. 2013). Habitat characters such as size, amount, orientation, position and structure reflect the complexity, dominance and heterogeneity of forest landscape (Garmendia et al. 2013). The complexity and dominance are promoting the nature of ecological process like survival capacity, specie colonization, movement, contrast, extinction, etc. Therefore, structural composition of the forest habitat has become a crucial subject in landscape ecological evaluation as well as wildlife management.

3.1.1.1 Habitat Fragmentation and Loss

The most urgent issue in biodiversity management is habitat fragmentation and loss that concerns with its structure and configuration. In South Bengal, forests become highly fragmented and isolated (Fig. 3.2).

Fragmentation is of two types such as structural and geographical fragmentation. Both fragmentations occurred due to encroachment and forest regeneration argued by Mandal and Chatterjee (2018). Mostly agricultural expansion by encroaching forest land makes the forest habitat fragmented structurally (Fig. 3.3). This type of fragmentation decreases the core and increases the edge habitat (Williams and Johansingh 1996; Dramstad et al. 1996; Mandal and Chatterjee 2019). As a result, existing wild metapopulation exposed and loses their normal movement. This wildlife behavioural pattern leads to wildlife conflict in this region (Mandal and Chatterjee 2020a, c).

As per State Forest Report (2016–2017), forest regeneration increased forest cover, but it isolates the forest patch regionally which impact on geographical fragmentation of forest patches in most of the western districts in South Bengal expect Sundarbans in south 24-Parganas. According to Mandal and Chatterjee (2021c) in Jhargarm district, the amount forest cover increased during past decades that attract wildlife but due to more isolated habitat (Fig. 3.4) it causes HWC so, geographical fragmentation is a very significant problem in South Bengal for managing wildlife as well as WHC (Santra

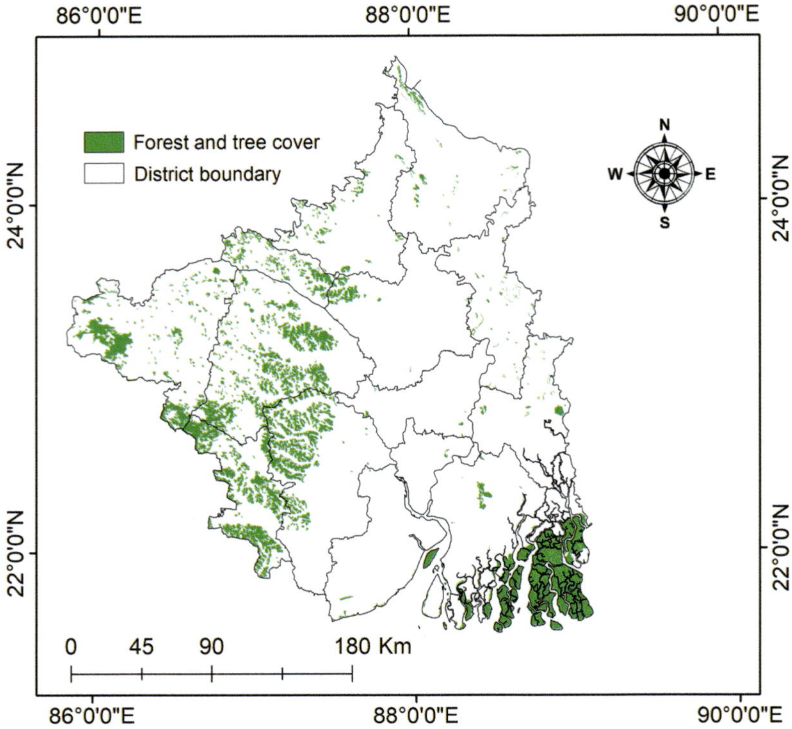

Fig. 3.2 Presents nature of forest fragmentation in southwestern districts of West Bengal State

2003). This situation happens not only in Jhargram district but also predominant in all the forested districts in West Bengal.

3.1.1.2 Habitat Edge Contrast

Forest habitat fragmentation leads edge contrast. Edge of the forest is simply the amount of perimeters of the forest patch. So the structural shape of the habitat determines its edge not the habitat size or amount. When a forest patch is structurally uniform shaped like square or circle, then edge will be minimum (McGarigal and Marks 1995). Deviation from uniformity indicates complexity and it means more amount of edge in a patch. From Figs. 3.2 and 3.4, it is very cleared that forest patch of South West Bengal becomes more complex. This is one of the causes of vulnerability of wildlife in the manmade landscapes (Chatterjee et al. 2014; Mandal and Chatterjee 2021b). High rate of crop raiding by elephant (Plate 3.1) and wild boar become a common phenomena in the forest edges (Chowdhury et al. 1997; Chanda 1996; Singh 2006; Mandal 2018; Mandal and Chatterjee 2020b).

3.1.2 Plant Species Composition

Plant species composition within a forest justifies the qualities of habitat that attracts or deters wildlife. Similarly, wildlife habitation pattern determined by the nature of plant species association and composition. Compositional characters of plant association play several fundamental roles for resting, hiding, foraging of wildlife animal in surroundings (Roberts and Gilliam 1995; Garibaldi and Turner 2004; Mandal and Chatterjee 2021a). Therefore, nature of plant composition and nature of response of wildlife are the important parameters for understanding the regional biodiversity. This relationship varies region to region and continent to continent depends on behavioural adaptation of wildlife.

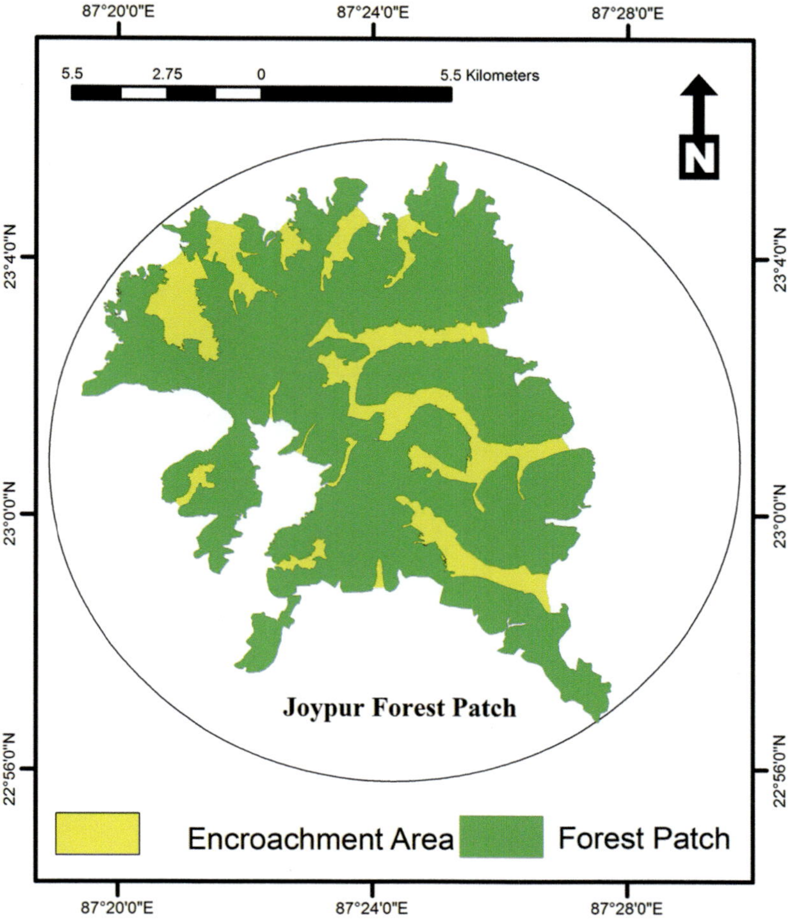

Fig. 3.3 Structural fragmentation and habitat loss in Joypur forest patch of Bankura district of West Bengal State

Plant species composition in South Bengal district forests except Sundarbsns in south 24-Parganas are inhomogeneous or diversified in nature. Not only the spatial variation but their composition varies in different forest patches (Kulandaivel 2010; Mandal and Chatterjee 2021c). Therefore, corresponding habitat quality in these districts is also different at micro-scale. Compositional characters of trees such like dominance, status, height, forest undergrowth are diverse in different forest that ultimately changes the forest quality. The coming chapter deals with this issue. It is found that similar type of plant species community generates different habitat conditions in different locations (Elzinga et al. 2001). One of the most important criteria is nature of microhabitat characters. Components like shadow, dominance or abundance, canopy, hiding place, ground cover, etc. are changed with changing plant species association and composition (Plate 3.2).

3.1.2.1 Plant Species Diversity and Density

Plant species diversity is the number of plants within a given geographical area that indicates richness. Another component of diversity is evenness, i.e. abundance distribution of species in a community. High plant species richness always is valuable for ecosystem balance. It has multi-functional effect in managing soil erosion, stable microclimate, safe shelter for wildlife, nutrition of the forest trees and maintaining food chain in an ecosystem. It is observed that, plant species diversity is different in south West Bengal and the Sundarbans forest, hence the forests are categorized into different types (Plate 3.2).

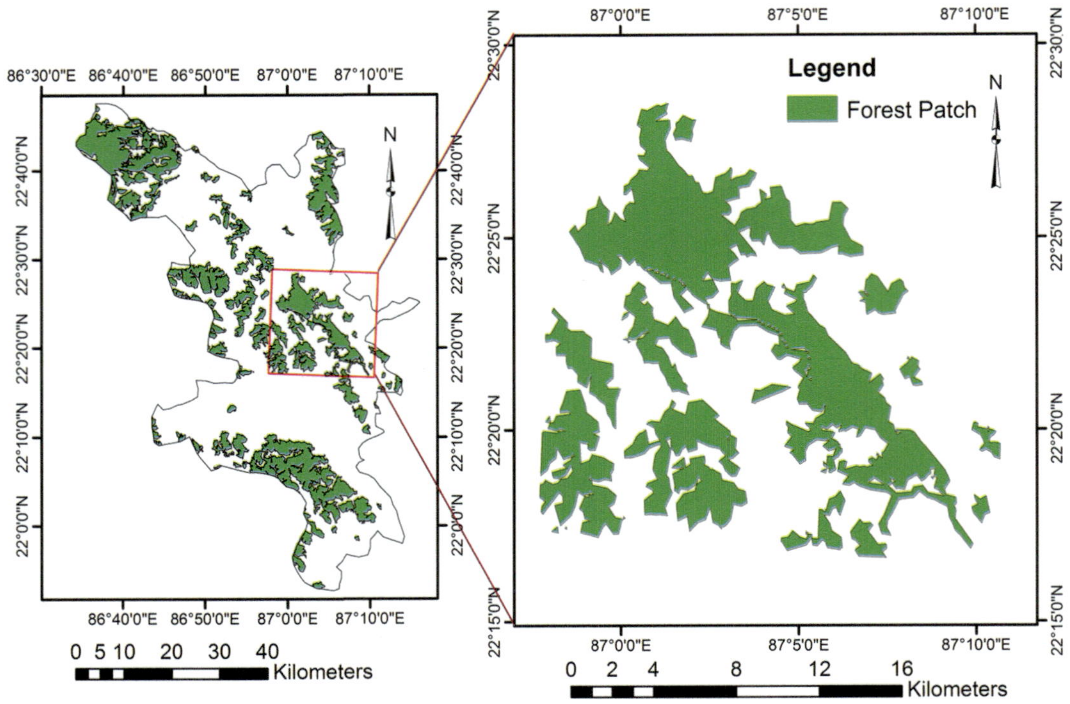

Fig. 3.4 Nature of habitat isolation or geographical fragmentation in Jhargram district

Plate 3.1 Elephant movement along the edge of forest and agricultural land at Moupal Beat in Paschim Medinipur Forest Division

Plate 3.2 Compositional characters of different forest habitats in Bankura district under South West Bengal. **a** Akashmoni-dominant forest with poor plant species diversity, **b** sal-dominant forest with moderate-to-high plant species diversity, **c** Eucalyptus-dominant forest with very poor plant species diversity and **d** mixed forest with dotted sal represents very high plant species diversity

Generally in the mature sal and mixed forest in Bankura district, a high-to-moderate diversity was found by Mandal and Chatterjee (2021a). The common trees diversity in these forests is listed in Table 3.1. It is observed that Eucalyptus and Akashmoni tree-based forest hold minimum diversity with very poor habitat condition for wildlife.

Another plant species compositional character is its density. Simply, plant species density is the mean inter-distance of given individual plant within a given space (Lal et al. 2019). Density measurement methods and techniques are elaborately explained in the next chapter of this book. Tropical dry deciduous forest in South Bengal is mostly manmade in nature. Therefore, density of the forest defers in different forest types and different areas (Kushwaha and Nandy 2012). It is observed that mature sal and mixed forest have better plant species density and diversity, which offers a favourable habitat for wildlife in south Bengal. Some Akashmoni-based forests hold better density, but lacking to generate forms good qualities of diversity. So, these forests cannot provide good habitat for wildlife like resting, hiding and foraging.

3.1.3 Locational Composition

After structural and species composition, locational composition has been taken into consideration for understanding connectedness and connectivity of wildlife in this region. Wildlife connectivity directly related to presence, position, aggregation and proximity of some landscape units. These are—forest as a shelter and food; water body as drinking source; agricultural crop land as a secondary food source and cultural or physical land units as barrier. Out of these,

Table 3.1 Common trees species diversity in sal and mixed forest in Bankura and Paschim Medinipur districts

Scientific name	Local name
Adina cordifolia	Korom/Holdu
Azadirachta indica	Neem
Buchanania lanzan	Piyal
Diospyros excelsa	Kendu
Cochlospermum religiosum	Golgoli/Galgal
Madhuca latifolia	Mahua
Shorea robusta	Sal
Holarrhena antidysenterica	Kurchi
Semecarpus anacardium	Vella
Soymida febrifuga	Rahara
Pterocarpus marsupium	Peasal
Schleichera trijuga	Kusum
Terminalia tomentosa	Asan

forest habitat locational composition is the first criteria for understanding the connectivity because it is the basic criteria.

3.1.3.1 Habitat Connectivity

Connectivity is defined as "the degree to which the landscape facilitates or impedes animal movement" (Taylor et al. 1993). Movement or dispersal of wildlife in forested matrix is an important landscape ecological event (Farina 2006). The nature of movement of large herbivores depends on specific criteria. These are habitat gap, inter-distance between corresponding habitat and obstacles between the habitats (Harrison and Fahrig 1995; Schumaker 1996). It is found from the existing literatures that more proximate habitat with less amount of habitat gap, local animal species usually avoids frequent and unexpected movement. This ecological behaviour is good for WHC. But when forest habitat becomes more isolated and more cultural land unit present in habitat gap, then HWC may be more frequent and become immeasurable.

Landscape of South Bengal districts is highly complex and isolated due to landuse alteration. In regional context, inter-distance of some aggregated forest patch becomes very high (Fig. 3.5). The newly generated forest (after 1980) increases the amount of forest which promotes better proximity of the habitat. As a result many wildlife animals in these district recolonized (Sukumar 2003; Singh 2006; Mandal and Chatterjee 2021a). But, this very forest regenerate program did not considered the habitat gap for forest plantation. That's why gap between habitat patches increased and it behaves as a buffer for animal movement and safe connectivity. Not only habitat gap but also other human-induced landuse like road, canal, etc. isolates the forest habitat. Railway line from Khargapur to Adra and NH-60, SH-2 and several district highways dissected many forest patches (Plate 3.3) in South Bengal. Huge traffic flow on these networks disturbed animal normal movement including their habitation on a regular basis (Singh et al. 2002; Mandal et al. 2015). To avoid these disturbances, wildlife animal especially elephant in this region changes their common route and follows the rare route where human colonies or crop fields are common.

Wildlife connectivity and HWC are interrelated ecological phenomena. Human death and property loss are usual when connectivity of the habitat is worst due to anthropogenic intervention within the wildlife habitat. Only human death or injuries are not the issue, but wild animal death also occurs due to these events (Mandal and Chatterjee 2021b). For example three elephants died in 2016 after being hit by a train in Southeastern Railway near Peardoba

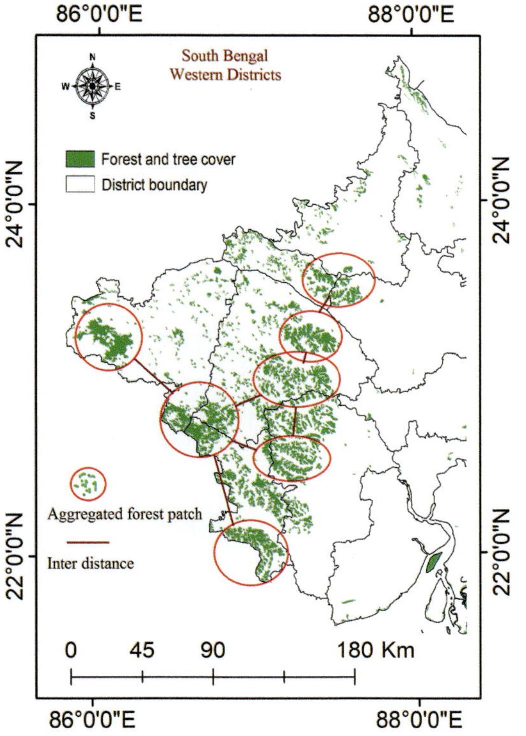

Fig. 3.5 Demarcation of habitat gaps and distance of some aggregated forest patches in western districts of South Bengal

forest that was rare case in South Bengal. Several report stated that both animal and wildlife deaths are increasing than the past in this region (Chatterjee 2016; Mandal and Chatterjee 2021d). Kangsabati canal crosses through several districts' forest patches which concretized in both sides (Fig. 3.3) crates a difficult situation for movement across the habitat to nearest habitat patch for wild animal in South Bengal.

3.2 Habitat Composition in Sundarbans Biosphere Reserve

Structural and locational habitat compositions are merely imperceptive in this deltaic Sundarban due to its high proximity. But characterising the species composition is one of the important indicator to understand wildlife ecosystem in this region. Having an unique mangrove ecosystem, Sundarbans represents structural habitat composition like fragmentation and connectedness. Habitat loss plays a significant role in this region. The foremost consideration is its species composition.

Indian Sundarbans has a magnificent ecosystem diversity which is more different from any other terrestrial ecosystem. Flora and fauna survives within the drastic environment in deltaic islands using their excellent adaptation capacity. A number of hydro-geomorphological changes have continued for shaping and reshaping the islands (Paul 2002, 2009) that make the estuarine habitat more complex. In this complex environment, ecosystem holds thousands of flora and fauna and they survive in its diversified habitat conditions (Plate 3.4) and niche. Biodiversity of Indian Sundarbans comprises plant composition, i.e. mangroves. Mangroves are defined as assemblages of salt tolerant trees and shrubs that grow in the intertidal regions of the tropical and subtropical coastlines. They grow luxuriantly in the places where freshwater mixes with seawater and where sediment is composed of accumulated deposits of mud (Tomlinson, 1986; Wadia 1961; Gopal and Chauhan 2006). Sundarbans (5366 km^2) supports thousands of species; these are 34 true mangrove plants species and 40 associated plant species, 250 species of fishes, 59 species of reptiles, 150 species of algae, 7 species of amphibian, 163 species of fungi, 32 species of lichen, around 200 species of birds, 39 species of mammals, besides numerous species of phytoplankton, zooplankton, benthos. (Chakraborty 2011).

Local people depends on Sundarbans for their livelihood and economy. They collect resources like fuel wood, timber, fishes, crabs and honey. For these they used to enter the functional prohibited zones. Construction of unscientific embankments in many areas of the Sundarbans, deforestation, frequent tourist visit, illegal reclamation of deltaic lands etc. creates barier for

Plate 3.3 Human-induced landscape as barrier for free wildlife movement. **a** Elephants crossing the road in Jhitka, Paschim Medinipur and **b** Kangsabati canal in Joypur forest in Panchet Forest Division under Bankura district, **c** Elephant crossing the railway near Basudebpur forest and **d** An Indian golden jackal stop crossing the SH-2 due to traffic at late evening

wildlife movement (Paul 2009; Chakraborty 2011). This alteration changes the composition of the mangrove forest and directly affects the fauna of the forest, especially the herbivorous.

3.3 Overall Outcomes

Geospatial information has a very important contribution in wildlife management. Spatial characters encompass a landscape. It contains three-dimensional characters with different types of patterns and processes. Some of these patterns are ecologically more significant. These patterns and processes regulate regional ecological system and also individual species-based ecological function. Ecological activities of wild animal like movement, colonization (Fig. 3.6), connectivity, extinction, etc. are directly connected with the nature of habitat composition like it structure, amount and plant species characters. Habitat function further depends on landscape character like structural configuration, amount, orientation, composition, etc. In a heterogeneous landscape, habitat patches are randomly distributed with varieties of floral composition which creates source–sink factors for animal habitat performance. It regulates nature of animal movement from one habitat patch to another patch. This is one of the major reasons for animal movement and colonization as well as HWC during their inter-patch movement. Ultimately ecological behaviour of an animal is controlled by landscape pattern and composition.

Plate 3.4 Species in different habitat conditions in Indian Sundarbans. **a** Dense mangrove in the edge of tidal river, **b** mangrove on the mudflat, **c** the legend tree in Sundarbans 'Sundari' (*Heritiera fomes*), **d** mangrove in tidal period, **e** spotted deer in the edge of mangrove forest and **f** crocodile in the mudflat

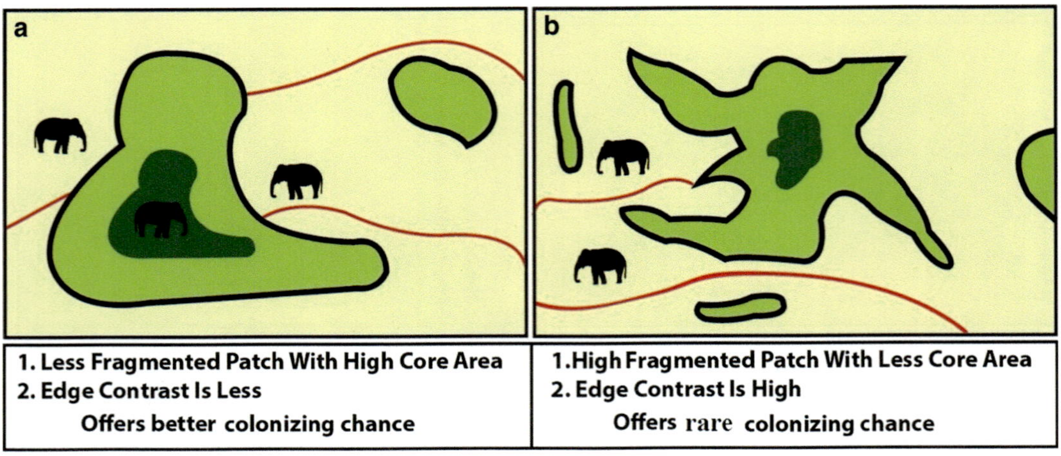

Fig. 3.6 Relation between forest habitat fragmentation and wild animal colonization on the basis of landscape ecological patterns and processes

References

Anon (2016–2017) State Forest Report. Forest Department, Govt. of West Bengal Directorate of Forests Office of The Principal Chief Conservator of Forests. Kolkata

Arroyo-Rodríguez V, Moreno CE, Galán-Acedo C (2017) La ecología del paisaje en México: logros, desafíos y oportunidades en las ciencias biológicas. Revista Mexicana De Biodiversidad 88:42–51

Chakraborty SK (2011) Mangrove ecosystem of Sundarbans, India: biodiversity, ecology, threats and conservation. In: Metras JN (ed) Mangroves: Ecology, Biology and Taxonomy. NOVA publisher, USA, pp 83–112

Chanda S (1996) Man-elephant conflict in South West Bengal. Joint Director of Sundarban Biosphere Reserve, West Bengal. IFS (1996 batch), 1–8

Chatterjee ND (2016) Man-Elephant conflict: a case study from Forests inWest Bengal. Springer, India

Chatterjee, N. Das, Chatterjee, S (2014) Changing Habitat and Elephant Migration from Dalma Wildlife Sanctuary, Jharkhand to Panchet Forest Division, Bankura, West Bengal: a biogeographical analysis. In: Climate change and biodiversity. Springer, pp 209–222

Chazdon RL, Guariguata MR (2016) Natural regeneration as a tool for large-scale forest restoration in the tropics: prospects and challenges. Biotropica 48 (6):716–730

Chowdhury S, Khalid MA, Roy M, Singh AK, Singh RR (1997) Management of elephant populations in West Bengal for mitigating man-elephant conflicts. Wildlife Institute of India, Dehradun

Cushman SA (2006) Effects of habitat loss and fragmentation on amphibians: a review and prospectus. Biol Cons 128(2):231–240

Cushman SA, Gutzweiler K, Evans JS, McGarigal K (2010) The gradient paradigm: a conceptual and analytical framework for landscape ecology. In: Spatial complexity, informatics, and wildlife conservation. Springer, Tokyo, pp 83–108

Das Gupta SP (1989) Forest eco-system in West Bengal. In: Budhadev C, Maiti A (ed) Forest and Forest Development in India. Inter-India Publications, D-17, Raja Garden Extn., New Delhi

Debnath D, Dasgupta S (2006) Livelihood generation and poverty reduction attempts in joint Forest Management activities in Madhya Pradesh. Int Rev 8(2):241–250

Dramstad W, Olson JD, Forman RTT (1996) Landscape ecology principles in landscape architecture and land-use planning. Island press

Dunning JB, Danielson BJ, Pulliam HR (1992) Ecological processes that affect populations in complex landscapes. Oikos 169–175

Elzinga CL, Salzer DW, Willoughby JW, Gibbs JP (2001) Monitoring plant and animal populations: a handbook for field biologists. Wiley

Fahrig L, Merriam G (1985) Habitat patch connectivity and population survival: ecological archives E066-008. Ecology 66(6):1762–1768

Farina A (2008) Principles and methods in landscape ecology: towards a science of the landscape (vol 3). Springer Science & Business Media.

Fernando P, Wikramanayake ED, Janaka HK, Jayasinghe LKA, Gunawardena M, Kotagama SW, Pastorini J (2008) Ranging behavior of the Asian elephant in Sri Lanka. Mammalian Biology-Zeitschrift Für Säugetierkunde 73(1):2–13

Farina A (2006) Principles and methods in landscape ecology. Springer, Landscape series

Forman RTT (1995) Land mosaic: the ecology of landscape and regions. Cambridge University Press, Cambridge, England

Garibaldi A, Turner N (2004) Cultural keystone species: implications for ecological conservation and restoration. Ecology Soc 9(3)

Garmendia A, Arroyo-Rodríguez V, Estrada A, Naranjo EJ, Stoner KE (2013) Landscape and patch attributes impacting medium-and large-sized terrestrial mammals in a fragmented rain forest. J Trop Ecol 29 (4):331–344

Gopal B, Chauhan M (2006) Biodiversity and its conservation in the Sundarban Mangrove Ecosystem. Aquatic Sci 68:338–354

Gupta R, Sharma LK (2019) The process-based forest growth model 3-PG for use in forest management: a review. Ecol Model 397:55–73

Gustafson EJ (1998) Quantifying landscape spatial pattern: what is the state of the art? Ecosystems 1(2): 143–156

Harrison S, Fahrig L (1995) Landscape pattern and population conservation. In: Mosaic landscapes and ecological processes. Springer, pp 293–308

Kotliar NB, Wiens JA (1990) Multiple scales of patchiness and patch structure: a hierarchical framework for the study of heterogeneity. Oikos 253–260

Kulandaivel S (2010) A paradigm shift in the elephant depredation in South Bengal. Divisional Forest Officer, Bankura North Division, Personal Communication

Kumar A, Marcot BG, Roy PS (2008) Spatial patterns and ecology of shifting forest landscapes in Garo Hills, India. In: Patterns and processes in forest landscapes. Springer, pp 125–139

Kushwaha SPS, Nandy S (2012) Species diversity and community structure in sal (Shorea robusta) forests of two different rainfall regimes in West Bengal. India. Biodiversity and Conservation 21(5):1215–1228

Lal HS, Ganguly S, Pramanik K, Prasanna PV, Ranjan V (2019) Plant diversity and vegetation structure in Sal (Shorea robusta Gaertn.) dominated forest of Dalma Wildlife Sanctuary, Jharkhand, India. Indian J Forest 42(1):83–90

Li H, Wu J (2007) Landscape pattern analysis: key issues and challenges. In: Key topics in landscape ecology. Cambridge University Press

Lovell ST, Johnston DM (2009) Designing landscapes for performance based on emerging principles in landscape ecology. Ecology Soc 14(1)

Lovett GM, Jones CG, Turner MG, Weathers KC (2005) Ecosystem function in heterogeneous landscapes. In: Ecosystem function in heterogeneous landscapes. Springer, New York, NY, pp 1–4

MacArthur RH, Wilson EO (1967) The theory of island biogeography: Princeton University Press, Princeton

Malhotra KC, Poffenberger M et al (1989) Forest regeneration through community protection. Proceedings of Working Group meeting on Forest Protection Communities, Calcutta

Mandal M (2018) Forest Range Wise Asian Elephant's (Elephas Maximus) Habitat suitability assessment through Food and Water Availability: a case study in Panchet Forest Division, Bankura, West Bengal

Mandal M, Chatterjee ND (2018) Quantification of habitat (forest) shape complexity through geo-spatial analysis: an ecological approach in Panchet forest division in Bankura, West Bengal. Asian J Environ Ecology 6:1–8

Mandal M, Chatterjee ND (2019) Forest core demarcation using geo-spatial techniques: a habitat management approach in Panchet Forest Division, Bankura, West Bengal. India. Asian J Geogr Res 2(2):1–8

Mandal M, Chattarjee ND (2020a) Land use alteration strategy to improve forest landscape structural quality in Radhanagar forest range under Bankura district. Eurasian J For Sci 8(1):1–10

Mandal M, Chattarjee ND (2020b) Geo-statistical analysis to understand nature of forest patch shape complexity in panchet forest division under Bankura district West Bengal. Indian J Ecology 47(1):96–101

Mandal M, Chatterjee ND (2020c) Elephant's habitat suitability assessment through geo spatial quantification in Panchet forest division, West Bengal. Ecofeminism Climate Change

Mandal M, Chatterjee ND (2021a) Geospatial approach-based delineation of elephant habitat suitability zones and its consequence in Mayurjharna Elephant Reserve, India. Environ Dev Sustain 23(12):17788–17809

Mandal M, Chatterjee ND (2021b) Forest landscape and its ecological quality: astepwise spatiotemporal evaluation through patch-matrix model in Jhargram District, West Bengal State India. Regl Sustain 2(2):164–176

Mandal M, Chattarjee ND (2021c) Estimation of forest ecosystem quality using GIS tool in Panchet forest division, West Bengal, India. In Forest Resources Resilience and Conflicts (pp. 203–213). Elsevier.

Mandal M, Chettarjee ND (2021d) Human-Elephant Conflict in Joypur Forest Influence Areas, West Bengal, India. Gajah 54:34–36

Mandal M, Chettarjee ND, Hazra J (2015) Elephant migration and colonization in Bankura district, West Bengal, India. Vidyasagar University. Indian J Geogr Environ 14:46–52

McGarigal K, Marks BJ (1995) FRAGSTATS: spatial pattern analysis program for quantifying landscape structure. Gen. Tech. Report PNW-GTR-351, USDA Forest Service, Pacific Northwest Research Station, Portland, OR

Merriam G (1984) Connectivity: a fundamental ecological characteristic of landscape pattern. In: Brandt J, Agger P (eds) Proceedings of the first international seminar on methodology in landscape ecological research and planning. International association for landscape ecology, Roskilde University Centre, Roskilde, Denmark, Oct 15–19, pp 5–15

Murcia C (1995) Edge effects in fragmented forests: implications for conservation. Trends Ecol Evol 10 (2):58–62

Naveh Z (1994) From biodiversity to ecodiversity: a landscape-ecology approach to conservation and restoration. Restor Ecol 2(3):180–189

O'Malley LSS (1908) Bengal District Gazetteers: Bankura. Bengal Secretariat Book Depot, Calcutta

O'Malley LSS (1911) Bengal District Gazetteers: Midnapore. Bengal Secretariat Book Depot, Calcutta

Palit S (1989) Present status of Forest Protection Committee; Forest regeneration through community protection: the West Bengal experience. In: Malhotra KC, Poffenberger MRK (eds) Proceedings of the working group meeting of FPC, Calcutta, 21–22 June 1989. West Bengal forest Department. pp 5–8

Palit S (1991) Participatory management of forests in West Bengal. Indian Forester 117(5):342–349

Paul AK (2002) Coastal Geomorphology and Environment. Published by- Ajoy Bhattacharya of acb publications, pp 1–575

Paul AK (2009) Man-Environment interactions in the single largest mangrove forest along the shoreline of the Bay of Bengal : a case study of Sundarbans. In: Sharma HS, Kole VS (eds) Geomorphology in India, pp 263–286

Pickett STA, Cadenasso ML (1995) Landscape ecology: spatial heterogeneity in ecological systems. Science 269(5222):331–334

Roberts MR, Gilliam FS (1995) Patterns and mechanisms of plant diversity in forested ecosystems: implications for forest management. Ecol Appl 5(4):969–977

Santra AK (2003) Studies on Elephant (*Elephas Maximus*) movement In South West Bangal And Its Impact. West Bengal University of Animal and Fishery Sciences, Kolkata

Schumaker NH (1996) Using landscape indices to predict habitat connectivity. Ecology 77:1210–1225

Singh AK (2006) Ecological Investigation of Human-Elephant Conflicts in South West Bengal. Saurashtra University

Singh AK, Singh RR, Chowdhury S (2002) Human-elephant conflicts in changed landscapes of south West Bengal. India. Indian For 128(10):1119–1132

Sitompul AF, Griffin CR, Rayl ND, Fuller TK (2013) Spatial and temporal habitat use of an Asian elephant in Sumatra. Animals 3(3):670–679

Sudhakar R, Raha AK (1994) Forest change detection study of nine districts of West Bengal through digital image processing of Indian Remote Sensing Satiate data between 1988 &1991–Procedural Manual and Inventory. Regional Remote Sensing Service Center, Kharagpur and Forest Department, Govt. of West Bengal Joint Collaborating Project

Sukumar R et al (2003) The living elephants: evolutionary ecology, behaviour, and conservation. Oxford University Press

Taylor PD, Fahrig L, Henein K, Merriam G (1993) Connectivity is a vital element of landscape structure. Oikos 571–573

Tomlinson PB (1986) The Botany of mangroves. Cambridge University Press, Cambridge, p 414

Wadia DN (1961) Geology of India. MC Muller & Co., London

Williams AC, Johansingh AJT (1996) Status survey of elephants, their habitats and assessment of elephant-human conflict in Garo hills, Meghalaya. Final report. Wildlife Institute of India, Dehradun

Habitat Quality Assessment Through Landscape Pattern, Plant Species Composition and Landscape Connectivity: Case Study from Bankura District

4

Abstract

The central theme of this chapter is to focus on the continuous interaction between pattern and process over the landscape. Pattern is formed by patches of landuse/land cover and the processes indicate the function of the landscape. Species richness depends on pattern of landscape. Landscape pattern analysis is always three dimensional in nature. Digital Elevation Model (DEM) is used to portray landscape pattern. The investigation of the present work is focused to analyze the elements of the landscape (habitat shape, core and edge) and quality of habitat. This chapter tries to investigate several landscape indices using statistical techniques and measure habitat quality and ecological functions at class and landscape level. The aim to this chapter is to identify suitable habitat for wildlife in the forest ranges.

4.1 Introduction

Fundamental landscape ecological studies address three major aspects: (i) spatial pattern of the presented land mosaic, (ii) compare these patterns with other landscapes and (iii) address obtainable ecological processes of the landscape. All these various aspects of landscape pattern assist to understand the dynamism of different habitats (Turner 1989, 1990; Turner et al. 1991), and it is defined by homogeneity, heterogeneity, shape regularity, etc. (Li and Reynolds 1995). The present chapter tries to discuss how the patterns (configuration of the landscape) manage the ecological processes like animal movement, corridor selection, colonization and animal extinction. These processes formatively trigger the ecological status of the existing habitat. In similar way, these pattern and processes are conditional to various compositional factors at micro (patch) and macro (landscape) level. Hence, this chapter carries out some reasonable methods to analyze the process response system of some individual species' activities in response to their habitat interaction from study area.

Another qualitative aspect is habitat plant species composition. It has a potential role for both species occurrence and their abundances in that habitat (Harper 1977). Vegetation composition and characters like its community, types, age and growth are the recognized factors for assessment of habitat quality or suitability (Shapiro et al. 2004; Mandal et al. 2021b). Specific compositions of plant in the habitat also influenced individual species habitat selection. Floral composition and structure are very important for understanding the ecological connection between species occurrence and habitat performance (Gottschalk et al. 2005; Austin 2007). So, plant species diversity, density, health, dominance, age and succession are important for the sensitive species colonization. It means, plant species

M. Mandal and N. Das Chatterjee, *Geo-Spatial Analysis of Forest Landscape for Wildlife Management*, GIScience and Geo-environmental Modelling, https://doi.org/10.1007/978-3-031-33606-5_4

characters control visibility, resting and foraging nature of animals within a forest reserve. Another important factor is biological metabolism of animal species that depends on the microclimatic condition in a forest habitat (Nowakowski et al. 2018). Reciprocally, macroclimate is regulated by existing plant species within the habitat (Lembrechts et al. 2019). Similarly, the seasonal variation of vegetation composition effects animal migration process. Animal migrates from one habitat to nearest habitat (finer scale) and even from one region to another region (larger scale) depending on source–sink factors (Mandal et al. 2015). These factors again changed with the changing nature of vegetation.

Habitat connectivity is also an important criterion for habitat assessment. It is defined as "the degree to which the landscape facilitates or impedes movement" (Taylor et al. 1993). Animal movement is a frequent natural ecological process. It is related to pull or push and source–sinks factors. Some landscape features create obstacle in species movement (Hoare 1999; McGarigal et al. 2009). Migratory animal species move from one patch to another habitat patch within the home range. This function determines species colonization, extinction, persistence, etc. Therefore, movement path or connectivity of corridor is a landscape element that plays a key role in species movement (Anderson 1991). Species movement and connectivity both are determined by individual's decision about the landscape. Specifically, frequent migratory animal species avoid road (Forman 1995, 2012; Forman et al. 2000) electric fencing, trench, canal, human settlement industries, etc., near the habitat patch or between two habitat patches. Therefore, some landscape elements enhance movement of species and some create obstacle or barrier to move them. The species movement investigation is very important for understanding landscape function whether it acts as barrier or encourage movements.

The background landscape matrix of Bankura district is dominated by agriculture. Almost 59% of the total land is practised for agriculture. The next dominant landuse is forest cover. Forest patch shape is highly modified and convoluted due to human intervention and different landuse encroachments (Mandal et al. 2018). In the study area, movement of elephant becomes an urgent ecological issue stated by Mandal et al. (2021a). The present chapter tries to quantify forest range-wise potential connectivity to understand the nature of movement, structural pattern of the forest patches for assessing the habitat quality and plant species composition variety and variability for measuring habitat performance of existing wildlife species through landscape matrices and GIS-based analysis.

4.2 Habitat Quality Assessment Techniques and Methods

Landuse and land cover features are unique and important criteria for landscape pattern analysis. For this purpose, the base map was taken from Indian Remote Sensing Centre (IRSC) for 2018 (Fig. 4.1). This categorical map was prepared and processed through different software for landscape pattern analysis.

This classified map was reclassified and then clipped and segregated for 28 forest ranges individually under Bankura district. From these reclassified data, only forest class is considered for analysis (Fig. 4.2). Some related secondary information was collected from district forest authority and opinion from individual ecological expert for understanding existing ecological pattern, processes and their function were collected.

4.2.1 Habitat Pattern Characteristics

The habitat pattern and structure within landscape have three functions. These are habitat dominance, habitat complexity and habitat dependency. Depending on these functions, ecologists address the issues like species movement, predation, migration behaviour and colonization of wildlife. The present chapter considers all the above mentioned parameters for measuring habitat quality. Each parameter composes different indices from landscape matrix.

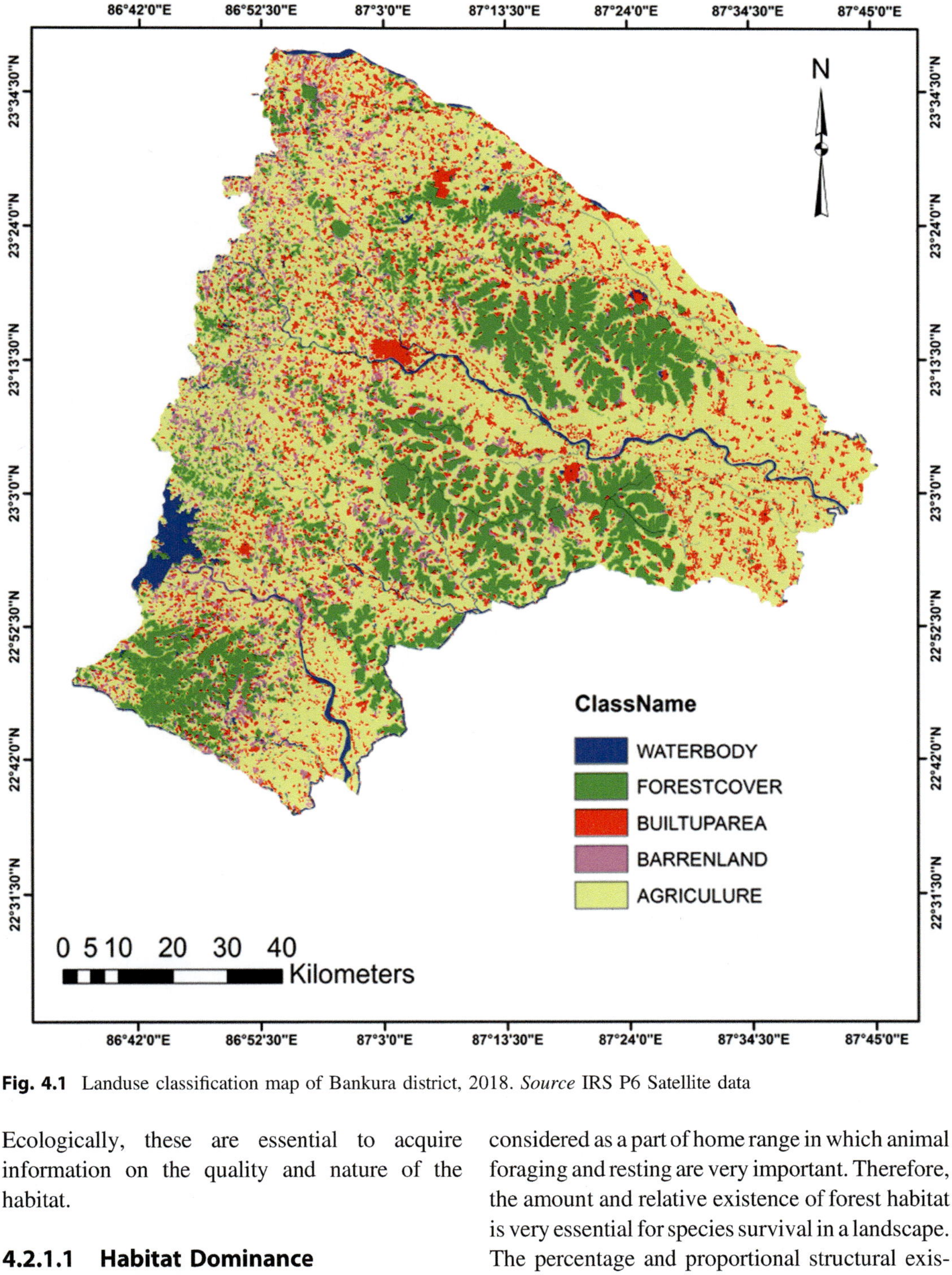

Fig. 4.1 Landuse classification map of Bankura district, 2018. *Source* IRS P6 Satellite data

Ecologically, these are essential to acquire information on the quality and nature of the habitat.

4.2.1.1 Habitat Dominance

Landscape is composed of several resource patches (Kulandaivel 2010). Vegetation cover is such a patch on which organisms find their 'habitat' (Farina 2008). For this reason, habitat patches are considered as a part of home range in which animal foraging and resting are very important. Therefore, the amount and relative existence of forest habitat is very essential for species survival in a landscape. The percentage and proportional structural existence of forest class determines habitat quality. Thus, this chapter considers LPI, PLAND and MPS indices for assessing habitat dominance of forest class over the total landscape (forest range).

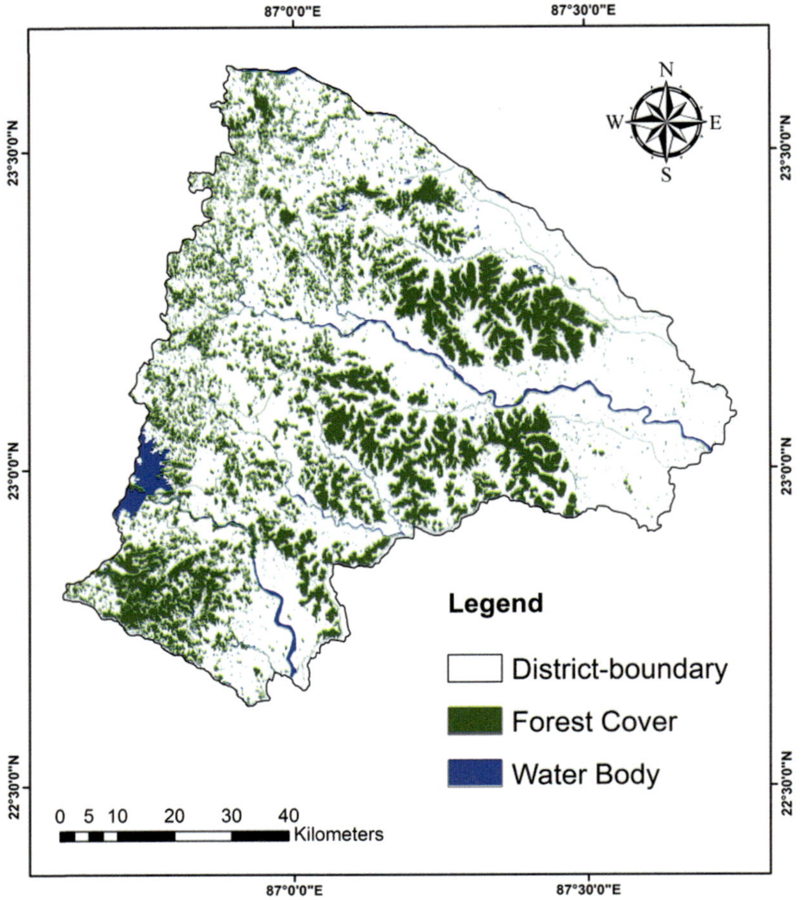

Fig. 4.2 Forest class map of Bankura district, 2018

McGarigal and Marks (1995)

1. Largest patch index (LPI)	
$LPI = \frac{\max_{j=1}^{n}(a_{ij})}{A}(100)$	a_{ij} = area (m^2) of patch ij class A = total landscape area (m^2)
Description	LPI equals the area (m^2) of the largest patch of the corresponding patch type divided by total landscape area (m^2) multiplied by 100 to convert to a percentage
Unit	%
Range	$0 < LPI \leq 100$ LPI approaches 0 when the largest patch of the corresponding patch type is increasingly small. LPI = 100 when the entire landscape consists of a single patch of the corresponding patch type
Comments	Largest patch index at the class level quantifies the percentage of the total landscape area comprised by the largest patch. It is simple measure of dominance

(continued)

2. Percentage of landscape PLAND	
$PLAND = \frac{\sum_{j=1}^{n} a_{ij}}{A}(100)$	a_{ij} = area (m^2) of patch ij class A = total landscape area (m^2) P_i = proportion of the landscape occupied by the patch type i
Description	PLAND equals the sum of the areas (m^2) of all patches of the corresponding path type dived by total landscape area (m^2) multiplied by 100 to change in percentage

(continued)

Units	%
Range	0 < PLAND ≤ 100 PLAND approaches 0 when the equivalent patch type becomes increasingly rare in the landscape. PLAND = 100 when the entire landscape consists of single path type
Comments	Percentage of landscape quantifies the proportional abundance of each patch type in the landscape. It is a measure of landscape composition and important in many ecological applications

3. Mean patch size (MPS)	
$MPS = \frac{\sum_{j=1}^{n} a_{ij}}{n_1}\left(\frac{1}{10{,}000}\right)$	a_{ij} = area (m^2) of patch ij class n_i = total number of patches in ij class
Description	MPS equals the sum of the areas (m^2) of all patches of the corresponding patch type, divided by the number of patches of the same type, divided by 10,000 (to convert to hectares)
Unit	Hectares
Range	MPS > 0, without limit
Comments	It is a measure of landscape dominancy and important in much ecological applications

4.2.1.2 Habitat Complexity

Habitat complexity refers to the heterogeneous nature of the landuse and land cover according to their structural shape. The structural complexity of the landscape generally interpreted through edge, shape and fractal dimension. Forest habitat edge, shape, fractal dimension, etc. plays vital role in the inter-patch processes (Mandal et al. 2019). Species migration, colonization (Buechner 1989) and animal foraging strategies (Mandal 2018) are depending on such factors (Forman and Godron 1986). Habitat edge complexity influences the movement of edge-specific species in a landscape. Shape and fractal dimension of the forest patch stand for the structural composition of the habitat. It influences interior species as well as native species diversity. Habitat indices like ED, MSI and AWMPFD are considered to measure the complexity.

McGarigal and Marks (1995)

1. Edge density	(habitat complexity)
$ED = \frac{\sum_{k=1}^{m} e_{ik}}{A}(10,000)$	e_{ik} = total length (m) of edge in the landscape involving patch type (class) i A = total landscape area (m^2)
Description	ED equals the sum of the lengths (m) of all edge segments involving the subsequent patch type divided by the total landscape area (m^2) multiplied by 10,000 to convert to hectares
Units	Metres per hectare
Range	ED ≥ 0 without limit ED = 0 when there is no class edge in the landscape
Comments	Edge density refers to edge length per unit area. It signifies comparative facilities available among landscape of varying size. It is the simple measure of class complexity
2. Mean shape index (MSI)	
$MSI = \frac{n_i \sum_{j=1}^{n} \frac{0.25 p_{ij}}{\sqrt{a_{ij}}}}{n_i}$	a_{ij} = area (m^2) of patch ij p_{ij} = perimeter (m) of patch ij n_i = number of patches in the landscape of patch type (class) i
Description	MSI equals sum of corresponding, {patch perimeter (m) divided by the square root of patch area (m^2) adjusted by a square standard} divided by number of patches in the landscape of patch type (class) i
Unit	None
Range	MSI ≥ 1 without limit MSI = 1 when the patch will be square and increases without limit as patches shape becomes more irregular
Comments	Mean shape index corrects the size problem of perimeter ratio index and it is the simplest and perhaps most straightforward measure of shape complexity

(continued)

3. Area-Weighted Mean Patch Fractal Dimension (AWMPFD)	
AWMPFD = $\sum_{j=1}^{n}\left[\left(\frac{2\ln 0.25p_{ij}}{\ln a_{ij}}\right)\left(\frac{a_{ij}}{\sum_{j=1}^{n} a_{ij}}\right)\right]$	a_{ij} = area (m^2) of patch ij p_{ij} = perimeter (m) of patch ij n = number of patches in the landscape of patch type (class) i
Description	AWMPFD equals the sum, across all patches of the equivalent patch type, of two times the logarithm of patch perimeter (m) divided by the logarithm of patch area (m^2), multiplied by the patch area (m^2) divided by total class area (sum of patch area for each patch of the equivalent patch type). (Li 1989)
Unit	None
Range	$1 \le \text{MPFD} \le 2$ AWMPFD approaches 1 for shapes with very simple perimeters such as circles or squares and approaches 2 for shapes with highly convoluted, plane-filling perimeters

4.2.1.3 Habitat Dependency

Habitat dependency is a function that is based on habitat core. Dependency usually improved when habitat patch becomes less fragmented, because fragmented forest generally holds less core area at specific edge depth stated by Watson et al. (2004). Therefore, on the basis of core area of forest class, which forest range is suitable for interior species and which is less suitable can be measured by applying edge buffering technique. At 300 m edge depth, total core area (TCA) can be measured by applying the following method. Moreover, core dependency can be measured using several indices such CAD, NCA, MCA and TCAI.

McGarigal and Marks (1995)

1. Total core area (TCA)	
$\text{TCA} = \sum_{j=1}^{n} a_{ij}^{c}\left[\frac{1}{10{,}000}\right]$	a_{ij}^{c} = core area (m^2) of patch ij based on specified edge depth (m)
Description	TCA equals the sum of the core areas of each patch (m^2) of the corresponding patch type, divided by 10,000 converts to hectares
Unit	Hectares
Range	TCA $\geq$ 0, without limit TCA = 0 when every location within each patch of the corresponding patch type is within the specified depth of edge distance from the patch perimeter. TCA is the same as the total class area when the edge depth distance is 0
Comments	Core area represents the area of a patch in the specified edge depth from patch perimeter
2. Core area density (CAD)	
$\text{CAD} = \frac{\sum_{j=1}^{n} a_{ij}^{c}}{A}$	a_{ij}^{c} = number of core area of patch ij based on specific-edge depth (m) A = total landscape area (m^2)
Description	CAD equals the sum of number of core area of patch ij based on specified-edge depth (m) divided by total landscape area (m^2)
Unit	None
Range	CAD $\geq$ 0, without limit CAD approaches 0 when core area of the corresponding patch type becomes increasingly rare in the landscape, because of increasing of smaller patches and/or more convoluted patch shape. A CAD value near the total landscape area when total

(continued)

	landscape is consists of a single patch and edge depth is 0
Comments	Core area density expresses number of core areas on per unit area basis that facilitates comparisons among landscape of varying size
3. Mean core area (MCA)	
$\mathrm{MCA} = \frac{\sum_{j=1}^{n} a_{ij}^{c}}{n_{ij}^{c}}(10,000)$	a_{ij}^{c} = core area (m^2) of patch *ij* based on specified-edge depth (m) n_{ij}^{c} = number of core areas in patch *ij* based on specified edge depth (m)
Description	MCA equals the sum of core area (m^2) of patch *ij* based on specified edge depth (m) divided by number of core areas in patch *ij* based on specified edge depth (m) multiplied by 10,000 to convert in hectares
Unit	Hectares
Range	MCA $\geq$ 0, without limit
Comments	Mean core area is the average core area of corresponding patch type in the landscape
4. Total core area index (TCAI)	
$\mathrm{TCAI} = \sum \frac{a_{ij}^{c}}{a_{ij}}(100)$	a_{ij}^{c} = core area (m^2) of patch *ij* based on specified-edge depth (m) a_{ij} = area (m^2) of patch *ij*
Description	TCAI equals the patch core area (m^2) divided by the total patch area (m^2), multiplied by 100 to convert in percentage
Unit	%
Range	$0 \leq 100$ When TCAI value is 0 the patch contains no core area at specified edge depth. TCAI value is 100—the patch contains maximum core area
Comments	Total core area index—a relative index that quantifies core area as a percentage of patch area

4.2.2 Plant Species Compositional Characteristics

A deep field investigation was conducted and various methods were applied for collecting plant species composition character. Lot of information were collected using DAFOR index. DAFOR stands for Density, Abundance, Frequency, Occasionality and Rarity of plant species. Most of the field investigations were conducted from 2017 to 2019 (Plate 4.1). Random sampling method was applied in selective forest patches which have larger core area at 100 m edge depth. A square grid of 50 m/50 m length is considered and it is divided into 10 m/10 m subgrids (Fig. 4.3). Grid information was collected from different locations of the forest patch. Plant species identification and count of the plant species were continued to get information about forest composition and quality of habitat. Number of trees in specific grid area was considered to get the density of a single grid. Every site is georeferenced by GPS. The information was taken in post-monsoon period, i.e. September–November of the respective year.

4.2.2.1 Plant Species Dominance

Dominance character of plant species represents habitat health. Sal (*Shorea robusta*) is dominant plant species in the forest patch of the entire region. Native or endemic plant species has a capacity to build stable ecosystem. After 1982, Acacia and Eucalyptus were planted extensively through Joint Forest Management scheme (JFM) (Chowdhury et al. 1997; Singh et al. 2002; Singh 2006). The native plant species are suitable for maintaining ecological balance than the introduced alien species like Acacia, Eucalyptus stated by several scholars (Das Gupta 1989; Pranovi et al. 2006; Foster et al. 2022).

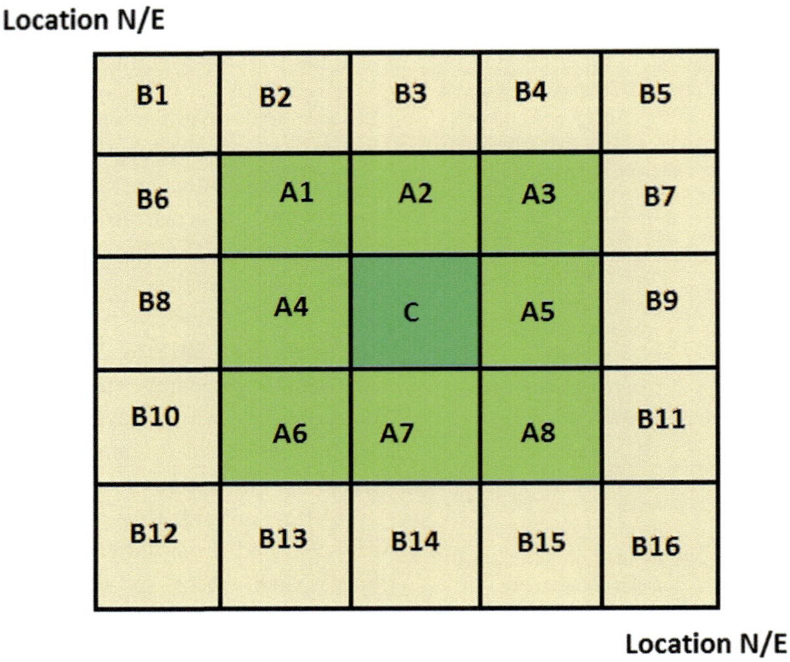

Plate 4.1 The grid matrix (25 subgrids) with 50 m/50 m length total grid area 2500 m^2. C, A and B are the grid Id

Fig. 4.3 Field survey photographs from Peardoba and Joypur forest under Panchet Forest Division, Bankura

1. Plant species dominance	
$\frac{\sum ABC^g}{\text{no. of individual tree count}}$	ABC^g = total grids of the matrix
Units	Number
Description	The highest number of individual tree is considered as first dominant. The next which is considered as second highest is second dominant and so on

4.2.2.2 Plant Species Density

Density of plant species has a great impact on health of forest habitat (Bera et al. 2022). Density refers the total number of plant species divided by target area. Highly dense forest habitat patches always support wilderness. Chance of local species extinction is less if the plant species density is high (Baltanas 1992).

2. Plant species density	
$\frac{\Sigma_{ABC^g} \text{ no. of trees count}}{A}$	ABC^g = total grids of the matrix A = area of the quadrat m^2
Units	Ratio
Description	When no. of trees per unit increase indicates high density and vice versa

4.2.2.3 Plant Species Health (DBH and Height of Tree)

Plant species health is recognized by its maturity level or age of the plant. In the study area, maximum forest is planted in nature (Anon 2006–2007). Ecologically, newly planted forest or regenerated forest is less significant than aged or old growth forest woodlots (Elzinga et al. 2001). Young sal forest has less qualitative than mature sal forest in the study area (Chatterjee 2016). To measure this concern, study has taken two factors. These are Diameter at Breast Height (DBH) and height of the tree following American Forests Tree-Measuring Guideline. Both the factors value when increased, the trend reflecting good health of the pant as well as forest habitat quality.

3. Plant species health or maturation	
1. $\frac{\sum_{ABC^g} \text{individual tree DBH}}{N_t}$ 2. Height of the trees by theodolite	ABC^g = total grids of the matrix N_t = total no. of the trees in the matrix DBH = Diameter at Breast Height
Units	Meter
Description	Both factors increasing value, increase maturation and health

4.2.2.4 Undergrowth

Forest undergrowth shows the surface coverage by grass, creeper, shrub etc. Maturity of the forest or level of ecological succession of the forest determines existing shrubs and creepers in the forest cover (Balfour and Bond 1993; Timilsina et al. 2007). Undergrowth character and condition are used as a available food for some herbivores (Sukumer 2003), as a resting place or hiding place for area-sensitive animal species (Eltringham and Malpas 1993). Further, it has a big role for microclimatic condition of the forest (Chen and Franklin 1990). The present work tries to measure undergrowth in percentage by a simple geometric method, i.e. summing up individual subgrid estimation.

4. Habitat under growth (up to 1.5 m height)	
$\frac{\sum_{ABC^g} \text{of individual } gca_{A1}}{A}(100)$	ABC^g = total grids of the matrix gca_{A1} = ground cover area of single grid in the matrix by eye estimation A = total area of all grids

(continued)

Units	Percentage
Description	When the value is 0, there is no undergrowth, and when 100, total grids are covered by undergrowth

4.2.2.5 Normalized Difference Vegetation Index (NDVI)

Greenness such green canopy cover reflects good health of forest. It gives better shadow and green leaves to the herbivore animals. NDVI is a reliable index to measure greenness (Gates, 1980). The differential reflection in the red and infrared (IR) bands allows density and intensity of green vegetation growth using the spectral reflectivity of solar radiation. Green leaves normally show better reflection in the near-infrared wavelength range than in visible wavelength ranges.

$$\text{NDVI} = \{(IR - R)/(IR + R)\}$$

source: FragState 4.2 Manual

- IR = pixel values from the infrared band.
- R = pixel values from the red band.

Low NDVI values indicate barren areas of rock, sand, or snow. Moderate values signify shrub and grassland, while high values specify mature greenness of forest. To justify the area of mature forest in the study area, only high NDVI value from 0.385219007 to 0.609375 has been considered as highly dense forest (Fig. 4.4).

4.2.3 Forest Habitat Connectivity

Animals move from one habitat patch to another similar habitat patch for their biological requirements such as shelter, resting, mating, food and hiding. Therefore, safe accessible movement path or corridors are always preferable by migratory animal species. Better habitat connectivity means undisturbed and shortest corridors offers stability in ecosystem of subpopulation present in the landscape. Thus, richness, extinction, availability and rate of colonization of local animal species strongly depend on habitat connectivity when habitat is isolated and proximate in modified landscape (Wilcove et al. 1986). Functional connectivity of landscape raises two questions. Which landscape element or land covers facilitates' movement and which landscape element creates hindrance to animal movement? Keeping these two questions in mind, the present study has focused to quantify range-wise connectivity for the entire landscape (forest range as total landscape) in Bankura district using specific landscape structural matrices.

4.2.3.1 Mean Nearest-Neighbourhood Distance (MNN)

(From McGarigal and Marks 1995)

Inter-habitat patch distance is an effective indicator that is associated with home range of migrating species. Different possibility may be faced by the migratory animals when they move one patch to another patch. Animal can easily move without any obstacle while distance is less. Therefore, connectivity of landscape (forest range) depends on nearest-neighbour distance.

$$\text{MNN} = \frac{\sum_{j=1}^{n} h_{ij}}{n_i}$$

h_{ij} = distance from it to nearest-neighbouring patch of the same type (class) i, based on patch edge-to-edge distance, n_i = number of patches in the landscape of patch type (class) i.

The value approaches 0 when the patches become more nearer and vice versa.

4.2.3.2 Mean Proximity Index (MPI)

(Gustafson and Parker 1992)

Habitat clustering pattern in a landscape determines species survival potentiality. The nature of isolation or dispersion of habitat indicates its connectivity. More disperse or isolated habitat patches offer poor quality in respect to species colonization, mobilization easy and safe foraging. It creates frequent disturbance and conflict.

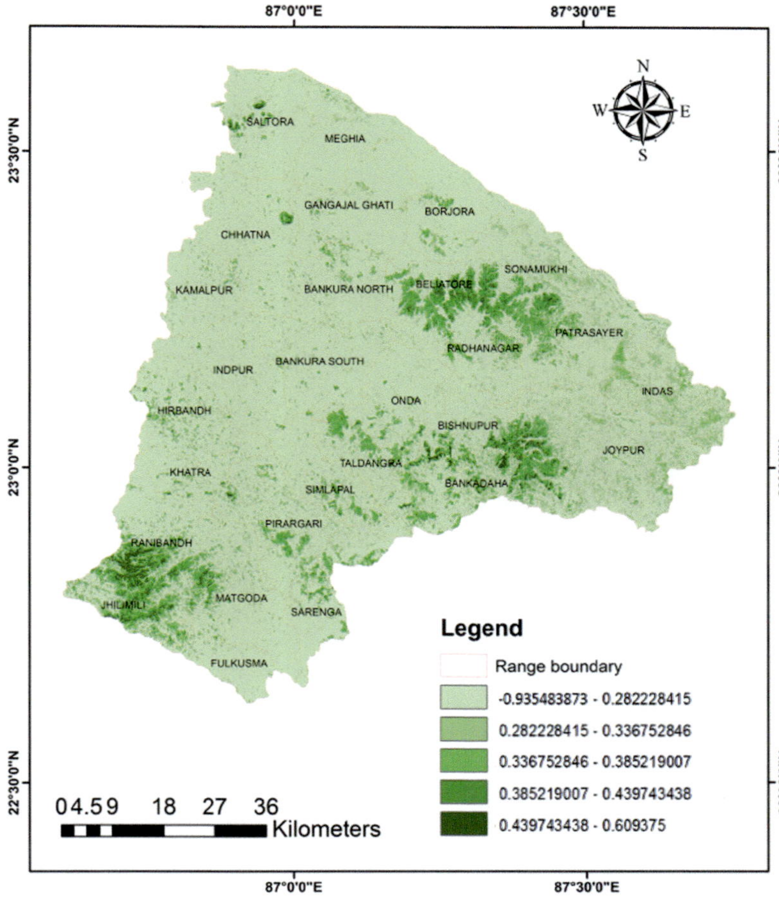

Fig. 4.4 Reclassified NDVI map of Bankura district. IRS- LISS-III ETM satellite data 2018

$$\mathrm{MPI} = \frac{\sum_{j=1}^{n}\left\{\sum_{s=1}^{n}\frac{a_{ijs}}{h_{ijs}^{2}}\right\}}{n_i}.$$

a_{ijs} = area (m^2) of patch ijs within specified neighbourhood (m) of the patch ij. h_{ijs} = distance (m) between patch ijs and patch ijs, based on patch edge-to-edge distance.

Index value is 0 when there is no similar habitat patch. Increasing value represents that habitat patches are closer and contiguous in nature.

4.2.3.3 Road Length Density

Human corridors like roads and railways are very common in a landscape matrix. These network lines are the symbols of development. But from ecological perspective, these landscape features make great obstruction to animal movement as well as regional ecology (Chatterjee and Chatterjee 2014; Mandal et al. 2021d). A transport line through the forest habitat never provides better movement condition.

$$\mathrm{RD} = \frac{\sum_{r=1}^{n} l_{ir}}{A}.$$

l_{ir} = length of the road into the forest in (m). A = total forest area of the landscape (individual forest range).

When value is high, the habitat is influenced by disturbance and when value is low, habitat become safe from disturbance.

4.2.3.4 Canal Length Density

Canal, natural or manmade interrupts safe animal mobility. It controls and changes the dispersal route of animal. Therefore, canal creates difficulties to the animal for their frequent movement from one place to another place.

$$\mathrm{CD} = \frac{\sum_{c=1}^{n} s_{ic}}{A}$$

s_{ic} = length of the canal into the forest in (m). A = total forest area of the landscape (individual forest range).

When the measuring value is high the range habitat is influenced by disturbance and lowering associability.

4.2.3.5 Built-up Area Density (Excluding Network Line)

Built-up unit in the landscape like settlement, industries or any type of construction over the landscape gives stress on ecological function. Especially, natural movement of animal dislocated by this specific land unit. It may be act as a source of noise, conflict, pollution, barrier, etc. For these reasons, animals avoid such composition (Woodward 2009) along their movement path. Therefore, dense built-up areas within a landscape reduce habitat suitability.

$$\mathrm{BAD} = \frac{\sum_{a=1}^{n} b_{ia}}{A}$$

b_{ia} = area of built-up patch i (m^2). A=total area of the landscape (individual forest range). The high value of BAD signify landscape is more disturbed and low value supports better animal existence.

4.2.4 Forest Range-Wise Distribution

After quantifying all the selected indices against 28 forest ranges (Table 4.1) it is observed that the difference of ecological indices are significantly related to the habitat quality. This work considered that ecological quality depends on the structural pattern and characteristics of the forest habitat (numbers, core area and edge). On the basis of these characters habitat complexity, habitat dominance and habitat dependency are quantified.

4.2.4.1 Habitat Pattern Quality Distribution

Bankura forest division consists of 28 forest ranges with different habitat qualities. Forest habitat patch is the second dominant landuse in the entire landscape (district) after agriculture. Species colonization processes are regulated by habitat dominance. Dominance indices' LPI value is found to be high in Patrasayer (14.25), Jhilimili (17.84) and Beliatore forest ranges, and these ranges are characterized by larger forest patches. These forest ranges may have capacity to hold maximum wild animal. Mejhia (0.27), Khatra, Fulkusma and Indas (0.02) forest ranges show very low dominance (Fig. 4.5). MPS (another dominance measuring index) shows that Patrasayer, Sonamukhi (47.31 ha), Radhanagar, Beliatore and Sarenga ranges have high dominance, whereas Khatra, Indpur, Kamalpur, Fulkusma, Mejhia, Indas, etc. (0.50 ha) have very poor dominance (Fig. 4.5).

Another dominance measuring index is PLAND or ZLAND. Using this index, high dominance is found in Radhanagar, Bankadaha, Bishnupur, Taldangra, Beliatore (36.08%), whereas Onda, Jhilimili, Sonamukhi, Fulkusma (2.04%), Indas (0.03%), Mejhia and Indpur have less dominance (Fig. 4.6). It is found that in case of NUMP, patch dominance is high in the western portion of the district (Fig. 4.6). Highest NUMP is found in Matgoda, Hirbandh, Pirargari and Saltora, but the patch size is too small. In case of percentage of forest habitat area, it is seen that Beliatore (36.08%), Radhanagar (28.35%), Bishnupur (27.66%), Bankadaha, Taldangra and Jhilimili ranges hold higher amount of habitable area than Indpur (3.71%), Mejhia and Indas (0.03%) (Fig. 4.6).

Habitat complexity totally depends on patch perimeter characters. The important index for measuring complexity is ED. High ED forest ranges are Jhilimili (47.51 m/ha), Bankadaha (36.64 m/ha), Radhanagar (30.23 m/ha), whereas low ED forest ranges are located at Indas (0.15 m/ha), Fulkusma (7.04 m/ha) and moderate ED found in Bishnupur (28.37 m/ha), Gangajal Ghati (23.30 m/ha) forest range area

Table 4.1 Forest range-wise individual ID and number of forest patches under Bankura district

ID	Name range	Nump	ID	Name of range	Nump
1	Saltora	302	15	Onda	241
2	Sonamukhi	130	16	Bankura South	283
3	Indas	13	17	Simlapal	310
4	Patrasayer	117	18	Kamalpur	384
5	Meghia	186	19	Indpur	290
6	Borjora	202	20	Hirbandh	570
7	Gangajal Ghati	295	21	Khatra	384
8	Chhatna	460	22	Sarenga	80
9	Bankura North	329	23	Pirargari	291
10	Joypur	142	24	Ranibandh	400
11	Radhanagar	135	25	Jhilmili	325
12	Bishnupur	143	26	Bankadaha	255
13	Taldangra	186	27	Matgoda	376
14	Beliatore	161	28	Fulkusma	107

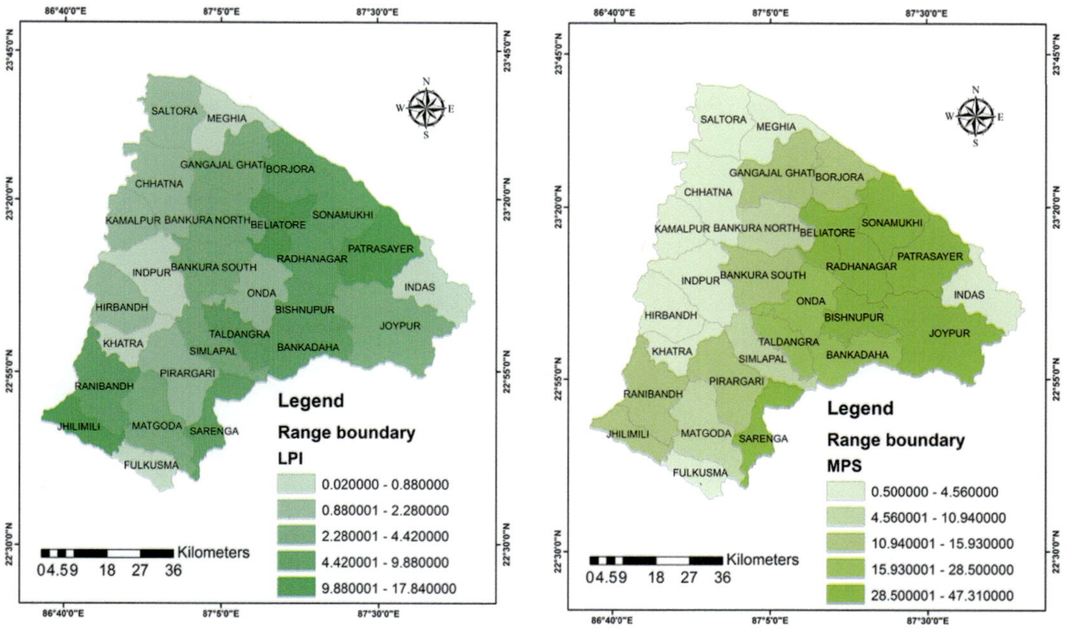

Fig. 4.5 Forest range-wise distribution of LPI and MPS for the representation of habitat dominance in Bankura district, West Bengal

(Fig. 4.7). Similar results also found in case of MSI. This index measures the encroachment rate. Ranges like Sarenga and Gangajal Ghati shows more irregular habitat shape, whereas moderate irregularity has found in Bankadaha, Bankura North, Bankura South, Onda, Joypur, Borjora, Jhilimili forest ranges' area. Less irregular shape is found in Indas forest range. Structural

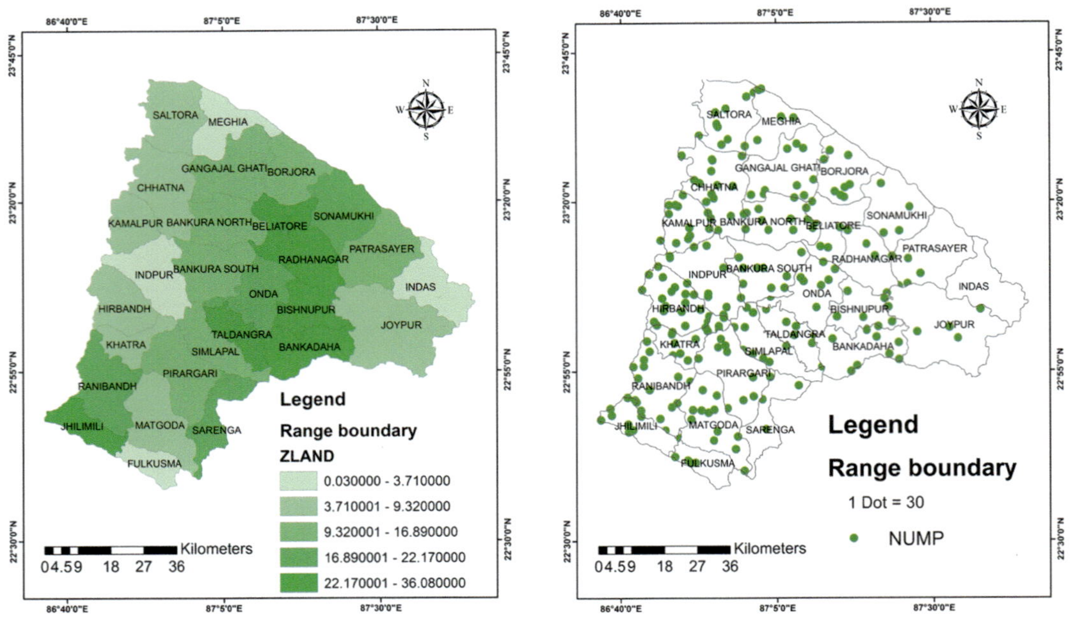

Fig. 4.6 Forest range-wise patch dominance on the basis of NUMP and PLAND, Bankura District

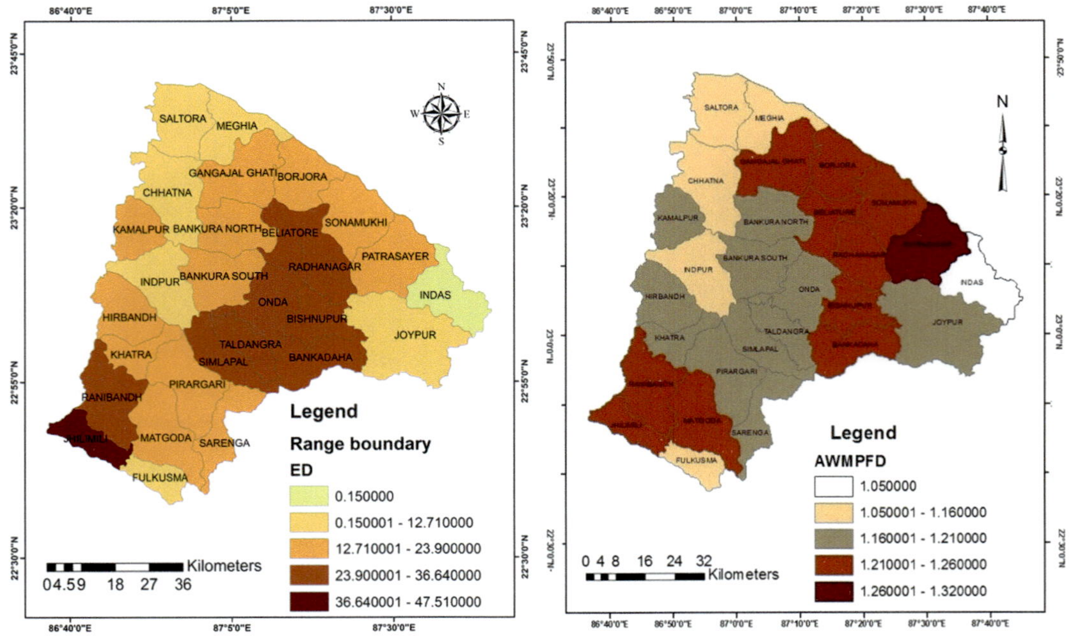

Fig. 4.7 Forest range-wise habitat complexity distribution on the basis of ED and AWMPFD, Bankura District

fragmentation of the forest patches is measured by AWMPFD, and the result shows that Patrasayer forest range is highly fragmented and Bankadaha, Bishnupur, Onda, Bankura North, Bankura South, Radhanagar, Sonamukhi, Borjora, Beliatore, Hirbandh forest ranges are moderately fragmented. Less fragmented forest range is Indas (Fig. 4.7) (Table 4.2).

Out of 28 forest ranges in Bankura forest division, five forest ranges have no core area at 300 m edge depth level. Sonamukhi forest range contains high core area of 881.75 ha. Using the dependency indices, it is found that the forest ranges located in the southeastern part of the district like Beliatore—804.5 ha, Ranibandh—835.25 ha and Joypur 709.5 ha have higher core area. Bankadaha, Bishnupur, Jhilimili, Radhanagar and Taldangra have moderate core area (529–705 ha). The forest ranges are like Saltora, Chhatna, Khatra and Indpur have less core area because of more isolated habitat patch (Fig. 4.8). Jhilimili, Beliatore, Onda, Simlapal, Sonamukhi and Ranibandh are ecologically more significant in case of CAD (Fig. 4.8). After getting the core area and number of core (Fig. 4.8), we used MCA index to measure which forest

Table 4.2 Forest range-wise value of selected ecological indices

FID	Name	MSI	AWMPFD	ED	MPS	LPI	NUMP	PLAND
0	Saltora	1.36	1.15	11.18	4.56	1.30	302	5.25
1	Sonamukhi	1.35	1.25	19.62	47.31	9.88	130	22.17
2	Indas	1.05	1.05	0.15	0.50	0.02	13	0.03
3	Patrasayer	1.33	1.32	18.49	42.12	14.25	117	16.01
4	Meghia	1.42	1.15	7.97	2.66	0.27	186	2.33
5	Borjora	1.38	1.25	19.15	15.49	6.16	202	13.62
6	Gangajal Ghati	1.59	1.23	23.30	12.43	2.84	295	12.08
7	Chhatna	1.35	1.15	12.71	4.07	1.34	460	5.63
8	Bankura North	1.47	1.19	21.90	8.15	3.50	329	11.34
9	Joypur	1.43	1.19	8.55	33.88	3.00	142	9.32
10	Radhanagar	1.39	1.26	30.23	42.82	9.42	135	28.35
11	Bishnupur	1.48	1.22	28.37	34.92	8.07	143	27.66
12	Taldangra	1.41	1.20	29.57	20.60	7.91	186	26.42
13	Beliatore	1.46	1.23	32.77	41.54	14.19	161	36.08
14	Onda	1.50	1.20	27.35	20.19	2.28	241	20.26
15	Bankura South	1.51	1.20	21.73	13.10	3.08	283	14.03
16	Simlapal	1.43	1.21	27.71	10.94	3.58	310	16.89
17	Kamalpur	1.36	1.18	19.46	2.63	1.45	384	5.87
18	Indpur	1.40	1.15	10.21	3.64	0.31	290	3.71
19	Hirbandh	1.37	1.18	23.90	3.15	1.09	570	7.76
20	Khatra	1.36	1.18	19.46	2.63	0.80	384	5.87
21	Sarenga	1.68	1.21	20.35	40.06	6.71	80	17.88
22	Pirargari	1.47	1.20	20.19	13.57	2.11	291	13.30
23	Ranibandh	1.39	1.24	28.69	15.93	5.94	400	21.96
24	Jhilimili	1.39	1.26	47.51	14.44	17.84	325	33.21
25	Bankadaha	1.55	1.23	36.64	28.50	7.29	255	29.85
26	Matgoda	1.38	1.23	17.12	7.15	4.42	376	8.91
27	Fulkusma	1.33	1.16	7.04	2.20	0.88	107	2.04

range have maximum number of cores and its proportion. Large number of core reduces the mean core area and vice versa when habitat area is constant. This condition is found in Saltora and Chhatna ranges due to presence of Susunia Hill in Chhatna and Biharinath Hill in Saltora forest range which have less structural patch complexity. Finally, TCAI have been measured. TCAI is the proportional abundance of habitat core area against total landscape (Forest range). The high proportion has been found in the middle and southeastern part (Fig. 4.8). The forest range areas of Joypur 14.75, Sonamukhi 14.34 and Ranibandh 13.11 show high value. Poor values are found in northwestern forest ranges like Kamalpur, Saltora, Indpur, Fulkusma, Hirbandh and Khatra under Bankura forest division (Figs. 4.8 and 4.9) (Table 4.3).

4.2.4.2 Plant Species Compositional Quality Distribution

Characteristics of plant species composition are very important for wildlife habitat suitability zoning (Ahmad et al. 2018). Rank and weighted method generally used for estimating forest habitat quality. Main components are density, openness and maturity of the forest. Present study adopted the same method to get forest range-wise habitat quality. A self-assessment rank-weighted method has been considered after Goparaju et al. (2018) and Mandal et al. (2021a). Preset study followed five-point scale and then weighted according to plant compositional character and function (Table 4.4). Then total weighted values are divided by total sampling number to get average weighted value for individual forest range. For example, sal species is dominant in the forest as the largest canopy cover, undergrowth, branching and ecosystem services. So it is more significant than Eucalyptus, Akashmoni and other plant species. For this reason, sal-dominant quadrate scored high weighted value in five-point scale and represents good quality (Table 4.5).

Similarly other four parameters, i.e. tree density, DBH, average tree height and ground cover values, are obtained from individual sampling sites from each forest range. The value of individual parameters under a sample site has been collected from different sampling plots. The total value of each parameter was then averaged for a sample site. After getting the total average value for each forest range these values have been assigned scores using five-point scale to get the habitat quality of each forest range.

After weighing the average sampled data, it is found that plant species compositional quality is better in Joypur forest under Bankadaha forest range (Fig. 4.11). Sonamukhi, Ranibandh, Jhilimili, Beliatore, Patrasayer and part of Joypur and Taldangra forest ranges hold better condition where 1 m inter-distance plant species have been recorded at the time of field survey. These ranges also covered with maximum mature sal and young sal-dominated forest habitat. Not only forest composition but the amount of forest cover is also better than other forest ranges. Therefore, plant species compositional value is high (out of 30) in these ranges like Bankadaha 21.8, Patrasayer 20.7, Ranibandh 20.6, Radhanagar 16.2, Sonamukhi 20.2 (Table 4.6). Except Ranibandh and Jhilimili, other high score ranges are clustered as neighbour. Lowest weighted values are found in the ranges like Mejhia, Indas and they have no significant forest cover also. For that reason, these ranges hold very poor quality. Onda 17.2, Gangajal Ghati 15.7, Bankura South 16.0, Fulkusma 15.0, Matgoda 13.3 scored moderate value with less-to-moderate habitat quality (Fig. 4.10) (Table 4.7).

4.2.4.3 Connectedness Quality Distribution

The study considered five parameters; these are proximity, nearest-neighbourhood distance, road density, canal density and built-up patch density for evaluating habitat connectivity. Out of five, two parameters proximity and nearest-neighbourhood distance determine the structural connectivity of the landscape and other measures the functional connectivity. Last three parameters act as a buffer or barrier for large size animal movement. Crossing the transport line or canal is very difficult for animal species (Chatterjee 2016; Mandal et al. 2021c). For this reason they always avoid this type of modified

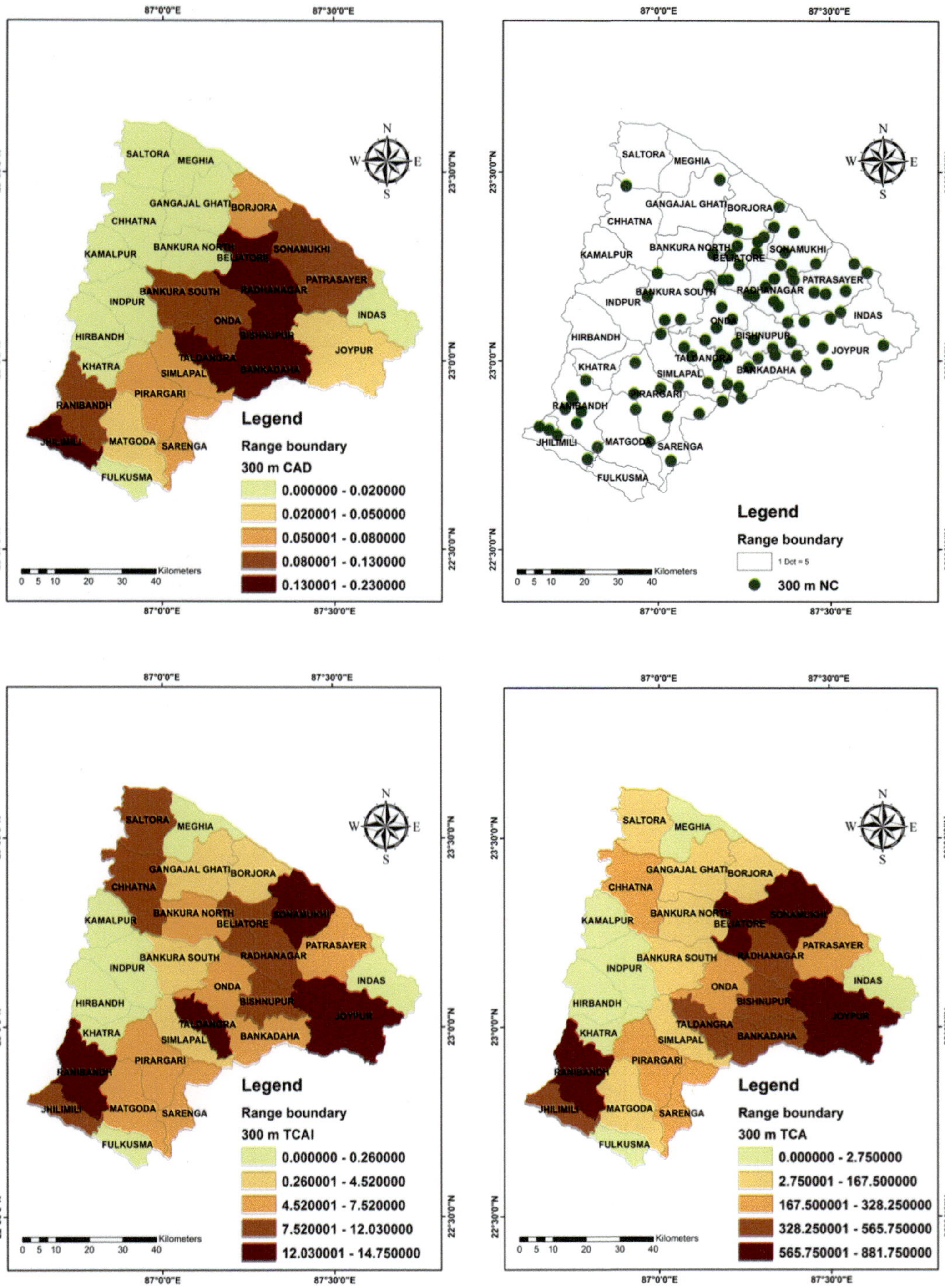

Fig. 4.8 Forest range-wise distribution of core area indices in Bankura district, West Bengal

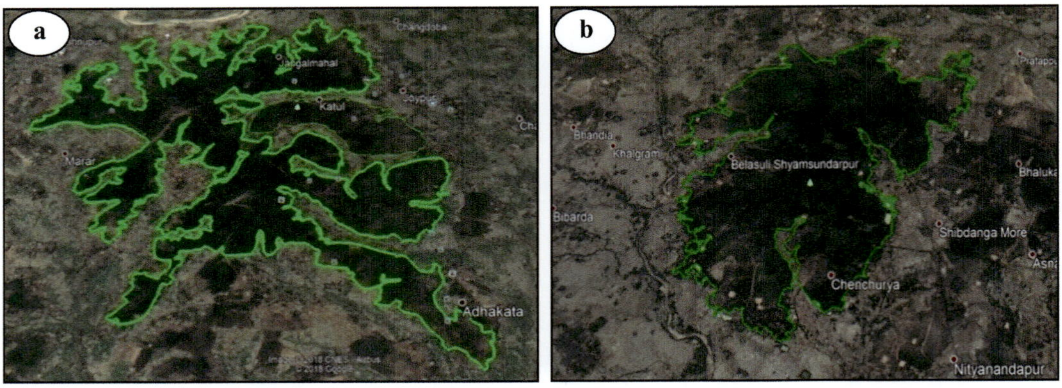

Fig. 4.9 Patch **a** has less amount of core area than patch **b** due to irregular shape, though total area of **a** and **b** patch are more or less equal. **a**—Joypur forest, **b**—Taldangra forest

Table 4.3 Forest range-wise value of core area indices at 300 m edge depth in Bankura district

FID	Name of forest range	NC	MCA (ha)	TCA (ha)	TCAI	CAD (m/ha)
0	Saltora	2	65	130	9.44	0.01
1	Sonamukhi	36	24.49	881.75	14.34	0.13
2	Indas	0	0	0	0	0
3	Patrasayer	32	10.27	328.25	6.67	0.1
4	Meghia	0	0	0	0	0
5	Borjora	16	6.05	96.75	3.09	0.07
6	Gangajal Ghati	5	16.25	81.25	2.21	0.02
7	Chhatna	4	56.19	224.75	11.99	0.01
8	Bankura North	5	33.5	167.5	6.25	0.02
9	Joypur	27	26.28	709.5	14.75	0.05
10	Radhanagar	33	17.14	565.75	9.79	0.16
11	Bishnupur	29	18.72	542.75	10.87	0.16
12	Taldangra	22	22.77	501	13.07	0.15
13	Beliatore	42	19.15	804.5	12.03	0.23
14	Onda	24	11.44	274.5	5.64	0.1
15	Bankura South	25	6.7	167.5	4.52	0.09
16	Simlapal	16	9.44	151	4.45	0.08
17	Kamalpur	0	0	0	0	0
18	Indpur	2	1.38	2.75	0.26	0.01
19	Hirbandh	2	0.5	1	0.06	0.01
20	Khatra	1	1	1	0.12	0.01
21	Sarenga	12	20.08	241	7.52	0.07
22	Pirargari	18	14.5	275.5	6.98	0.06
23	Ranibandh	29	28.8	835.25	13.11	0.1
24	Jhilimili	24	20.65	495.5	10.56	0.17
25	Bankadaha	42	10.05	422	5.81	0.17
26	Matgoda	10	14.95	149.5	5.56	0.03
27	Fulkusma	0	0	0	0	0

Table 4.4 Weighted score against individual parameter in a sample site matrix

FR	Sample no.	Dominant tree	Tree density (mean inter-distance) in m	Average tree DBH in inch	Average tree height in m	Ground cover %	Parameter weighted score	Total	NDVI area weighted value	Forest range weighted value
Sonamukhi	A	5	2	3	4	5	19	81/5 = 16.2	4	16.2 + 4 = 20.2 out of 30
	B	5	3	3	4	2	17			
	C	5	1	2	1	5	14			
	D	2	3	4	2	2	13			
	E	3	5	4	1	5	18			

landscape features. Therefore, connectivity for frequent movement of animal is restricted in disturbed landscape (Forman 1995; Mandal et al. 2018).

Forest range-wise connectedness is quantified using the indices like MNN in m and MPI. Connectedness, i.e. physical distance between forest patches, in the landscape. Here, connectedness signifies the connectivity of that population like elephants—*Elephas maximus* movement connectivity from one forest patch to another corresponding patch. The result of MNN shows lowest mean distance in middle-western part of the district. Forest ranges like Bankadaha, Kamalpur, Simlapal, Hirbandh, and Jhilimili, their inter habitat patch distance is about 100–200 m. Moderate average distance, i.e. 200–300 m is found in the forest ranges like Chhatna, Pirargari, Ranibandh, Khatra, Bankura South and Beliatore. Less connectedness is found in Onda, Sarenga, Bishnupur, Mejhia and Bankura North forest ranges, i.e. 300–400 m. Very low connectedness, i.e. more than 500 m, is found in the forest ranges of Indas and Joypur, Sonamukhi and Patrasayer (Fig. 4.11). The value of MPI shows high connectedness in the forest ranges like Sonamukhi (30866.65), Patrasayer (17819.61), Beliatore (12292.04), Radhanagar (6871.10), Jhilimili (10089.01), Ranibandh (11349.36), Bankadaha (7777.42) and Bishnupur (8308.77) correspondingly and their adjacent areas. Moderate MPI is found in Joypur (4525.17) and Borjora (2874.83). The index value is low to very low in the maximum ranges of northwestern portion of the Bankura district such as Indas (0.07), Chhatna (81.78), Indpur (104.35), Saltora (152.02), Khatra (158.24), Kamlpur (161.75), Mejhia (364.49), Bankura North (414.34), Simlapal (1621.98), Gangajal Ghati (674.82) (Fig. 4.11).

Connectivity processes are very sensitive due to the influence of disturbance and corridors between forest patches (Mandal et al. 2020b). Disturbances may be created by anthropogenic landuse like network line and canal. This landuse acts as a barrier or buffer for animal's frequent movement. In the study area, three landuse land

Table 4.5 Parameter-wise weighted value in different sample sites for Sonamukhi forest range under Bankura forest division

Dominant tree	Weighted value	Tree density (mean inter-distance) in m	Average tree DBH in inch	Average tree height in m	Ground cover %	Range habitat quality on the basis of plant species composition
Sal	5	Highest value – lowest value = the differentiate value has been classified into five class and weighted in five-point scale according to their function				Sum of all weighted values against every parameter divided by sample number
Other or Mixed	4					
Young sal	3					
Sonajhuri	2					
Pals/other	1					

Table 4.6 Forest range-wise total weighted value against selected parameters

FID	Range name	Weighted value out of 30 (against six parameters)
0	Saltora	16
1	Sonamukhi	20.2
2	Indas	0
3	Patrasayer	20.7
4	Meghia	0
5	Borjora	17.21
6	Gangajal Ghati	15.7
7	Chhatna	16.3
8	Bankura North	18
9	Joypur	18
10	Radhanagar	21.2
11	Bishnupur	20
12	Taldangra	20.2
13	Beliatore	19.7
14	Onda	17.2
15	Bankura South	16
16	Simlapal	17.6
17	Kamalpur	16
18	Indpur	15.7
19	Hirbandh	15.3
20	Khatra	16.6
21	Sarenga	17.6
22	Pirargari	15.6
23	Ranibandh	20.6
24	Jhilimili	20.2
25	Bankadaha	21.8
26	Matgoda	13.3
27	Fulkusma	15

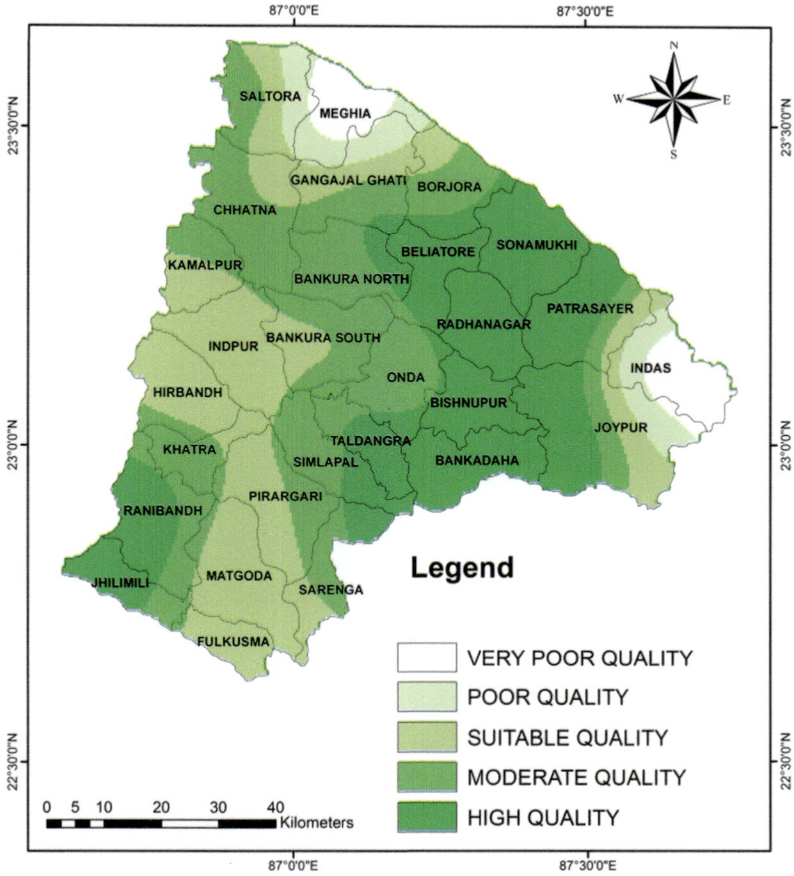

Fig. 4.10 Forest range-wise habitat quality map on the basis of species composition, Bankura district

cover types are measured as disturbance for animal safe movement.

Transport communication line and road network are randomly distributed in all 28 forest ranges of the district. Dense network lines are identified correspondingly in Bishnupur (3.60 m/ha), Bankura North (3.24 m/ha), Mejhia (3.10 m/ha), Onda and (3.12 m/ha) Sonamukhi (2.59 m/ha). Modest traffic line density is found in the forest ranges located at the middle part of the district. Relatively less dense traffic lines are found in southeastern forest ranges of the district which is ecologically suitable for safe movement of animal (Fig. 4.12). In case of canal density, high dense is observed in the forest ranges like Khatra, Kamalpur and Borjora and low density is found in northern portion of the district forest range like Gangajal Ghati, Saltora, Mejhia, Radhanagar and Beliatore (Fig. 4.12). Similarly, built-up patch density found high in Sarenga, Fulkusma and moderate density is found in Simlapal, Bankadaha, Indpur and Indas. Low density is found in northwestern part of the district which is significant for animal's frequent movement.

Generally, large amount of forest areas provide basic requirements for large animal population (Mandal et al. 2021d). In same way, greater patch area with high road and canal length have become more disturbed due to more frequent penetration of human, natural or human-made corridors inside it. From Fig. 5.9, it is clear that the forest ranges with high amount of forest area have also higher canal and road length and hence are unfavourable for species movement. Human–animal conflict is also common in these forest ranges like Bankadaha, Radhanagar, Sonamukhi, Beliatore and Bishnupur.

In Ranibandh and Jhilimili, the habitat patch is more clustered with less amount of habitat gap (Fig. 4.13). Road, canal and built-up areas are

Table 4.7 Range-wise forest habitat quality with some real photographs from different sampling sites

Forest Range Name	Quality
INDAS, MEJIA	  **Very Poorly suitable**
MATGODA, FULKUSMA, HIRBANDH, PIRARGARI, GANAJAL GHATI, INDPUR BANKURA SOUTH, KAMALPUR, SALTORA, CHHATNA	  **Poorly suitable**
KHATRA, ONDA, BORJORA, SIMLAPAL, SARENGA, BANKURA NORTH, JOYPUR	  **Suitable**
BELIATORE, BISHNUPUR, JHILIMILI, TALDANGRA, SONAMUKHI, RANIBANDH PATRASAYER	  **Moderately suitable**
RADHANAGAR BANKADAHA	  **Highly suitable**

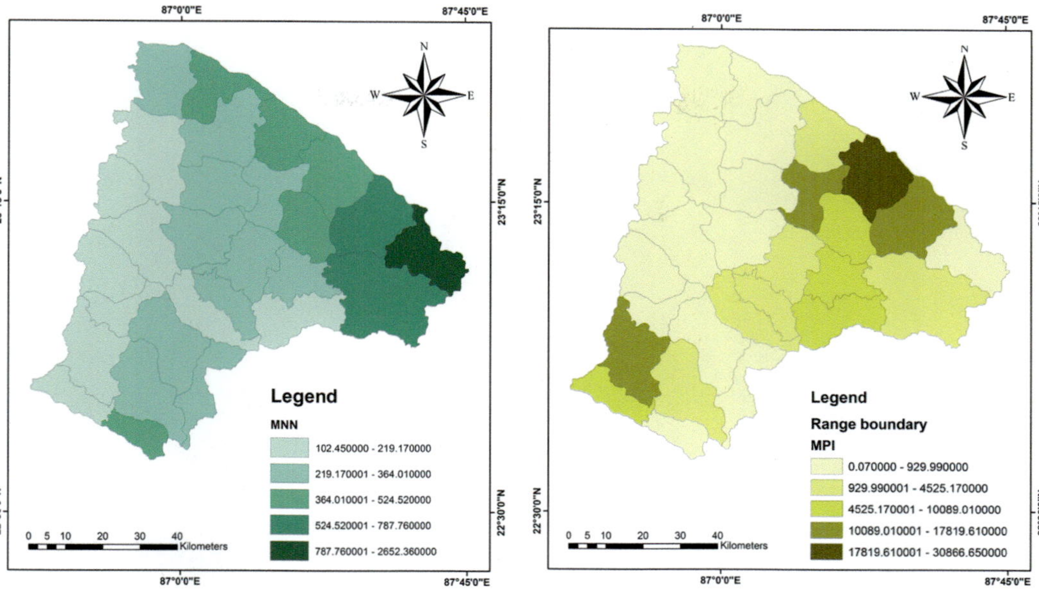

Fig. 4.11 Forest range-wise connectivity distribution by the indices of MPI and MNN under Bankura district

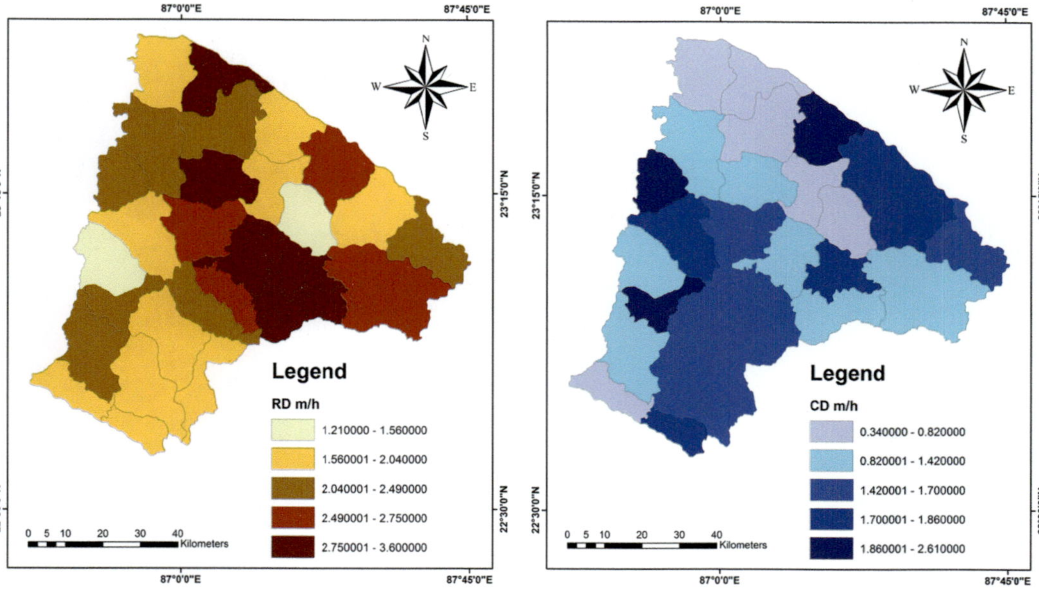

Fig. 4.12 Forest range-wise road density (RD) and canal density (CD) maps of Bankura district

less in amount. Therefore, connectivity of animal will be better with less hindrances. But the condition is just reverse in Fig. 4.14, i.e. the part of Bankadaha, Simlapal, Taldangra and Onda forest range where major roads like SH-2, NH-60, South-eastern railway line from Khargapur to Adra and Kansabati canal passes through the several forest patches. Habitat gap is also high in these forest ranges. Settlement units and agricultural lands present between these habitat gaps more frequently. Therefore, disturbance is high in these forest areas which cause more conflicts.

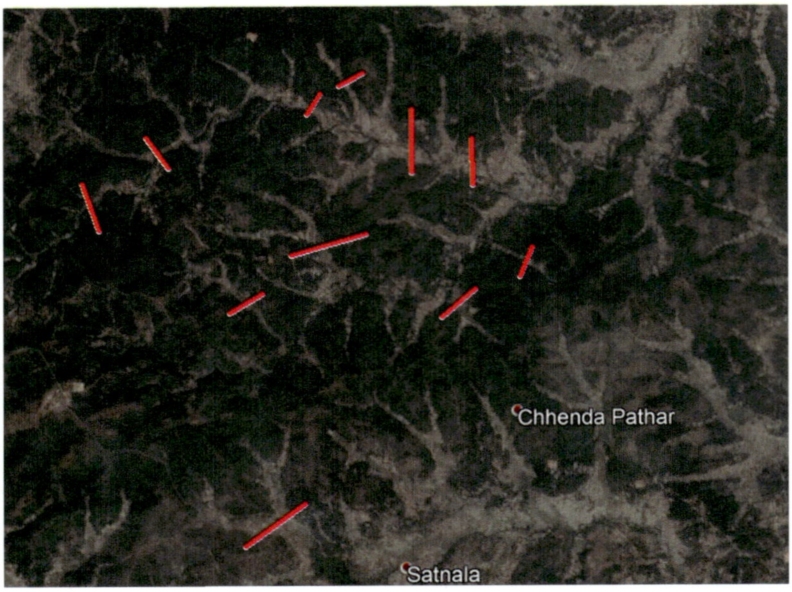

Fig. 4.13 Some part of true image of forest cover of Ranibandh and Jhilimili forest Range, Bankura District. *Source* Google image 2020

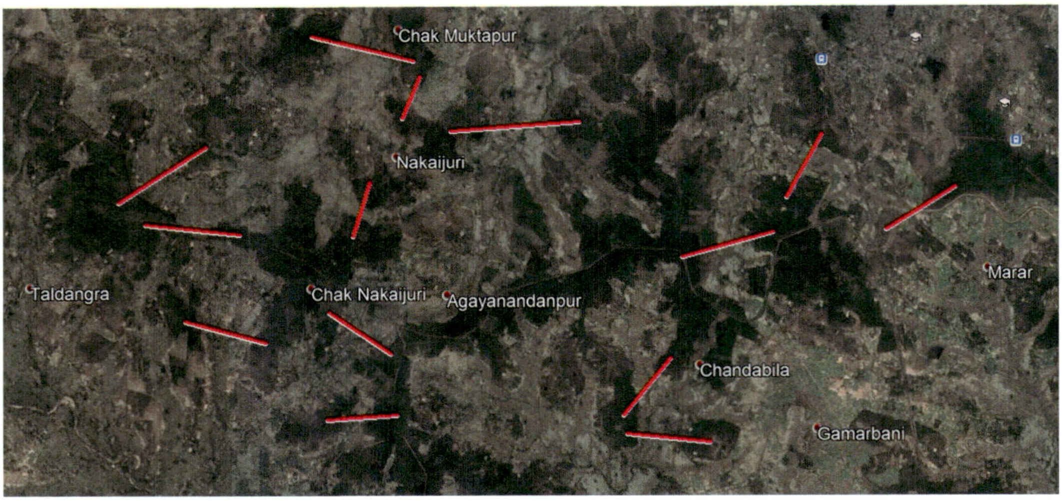

Fig. 4.14 Some part of Bankadaha, Simlapal, Taldangra and Onda forest ranges. The red lines point out the gap between two forest patches. *Source* Google image 2020

In case of built-up density, most of the southwestern forest ranges cover high density. For that reason, wildlife conflict intensity will more in these forest ranges. On the other hand the forest ranges like Joypur, Bishnupur-I and II, Radhanagar, Beliatore, Borjora, etc. cover less built-up density (Fig. 4.15). Theoretically, these conditions effects of human–wildlife conflict. But practically, it has been found that the incident of Human-wildlife conflict is more frequent in these forest ranges. Not only a single factor is responsible behind this phenomena, but many other landuse factors are also responsible. One important factor is high

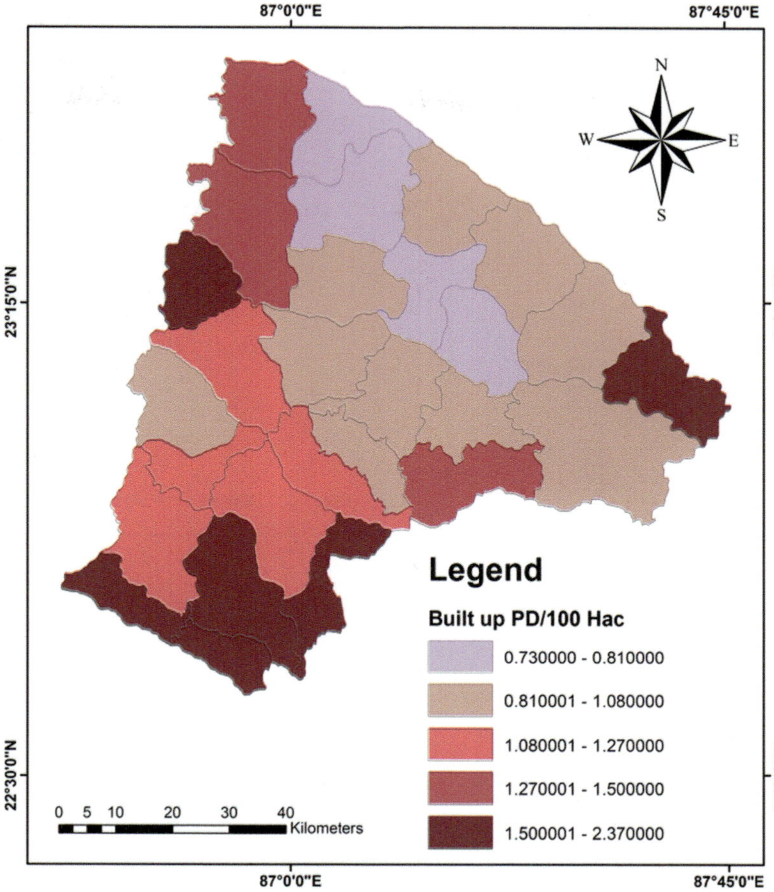

Fig. 4.15 Forest range-wise built-up patch density distribution map of Bankura district

density of in these forest ranges. Therefore, wild migratory animal like elephant usually take rest in these forest patch and invades the surrounding forest patches for their biological needs. As a result, these forest ranges are very prone to high conflict though they have more disperse built-up units.

4.3 Outcomes

After getting the results of different indices, it is found that the forest ranges in eastern part of the district (except Indas range) are ecologically more suitable than the western part. The result of the present study in this chapter displays the present ecological phenomena in different ways. It may be concluded that, highly suitable forest ranges of Bankura district are facing massive intrusion of migrated elephants (*Elephas maximus*). It is interesting to know that some residential bull elephants are appeared frequently in Joypur, Bankadaha, Sonamukhi, Borjora and Patrasayer. Wild boar, Hayne, Wild cat and some other indigenous animal species are also found in these forest ranges due to improved habitat condition. Spotted Deer conservation program started in the improves forest habitat in Joypur forest area. Jhilimili and Ranibandh forest ranges located in southwestern part of the district have good habitat condition. As a result, these two ranges are taken as a part of Mayurjharna Elephant Reserve which is located in the part of adjoining Purulia and Paschim Medinipur district (Fig. 4.16). Rare wild species Python and peacock are occasionally found here.

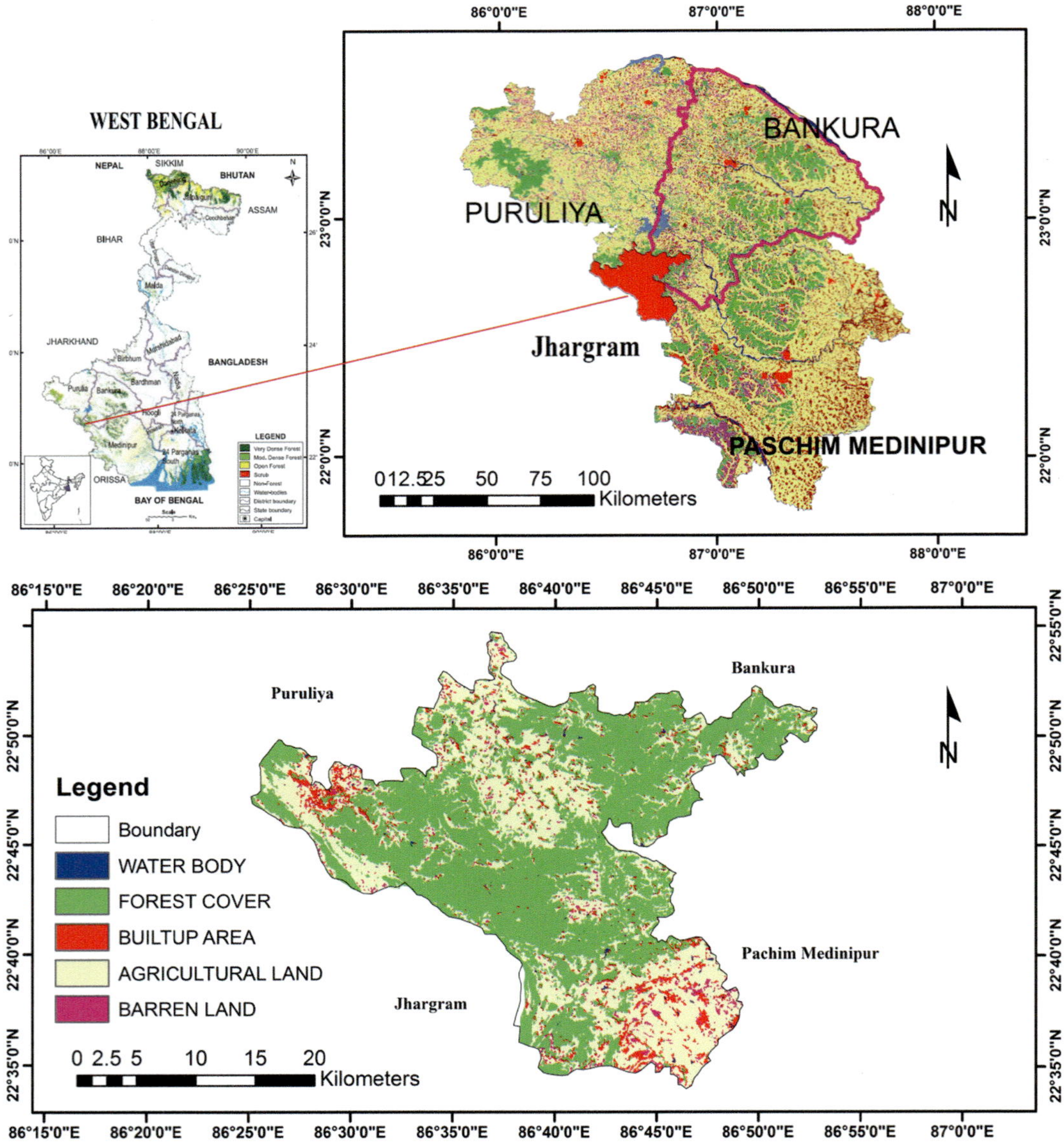

Fig. 4.16 Mayurjharna Elephant Reserve consisting of the areas of the districts Bankura, Jhargram and Puruliya in South Bengal. It was planned to address the issue of Human Elephant Conflict and elephant conservation

Most of the forest habitat edges in Bankura are fragmented in nature. The forest patch edge is surrounded by agricultural land, especially paddy and vegetables in south and southeastern forest ranges. For this reason, Bankadaha, Joypur, Onda, Simlapal, Patrasayer, Radhanagar, Bishnupur and Borjora forest ranges provide food for herbivores. Therefore, more edge contrast signifies conflict between human and animal, is an urgent problem in these forest ranges argued by Mandal et al. (2020c). The study reveals that the natures of forest habitat in these ranges become more fragmented and isolated. Thus, animals regularly face problem from different kinds of disturbances. Habitat gap and human intervention into the forest forced them to migrate from one forest patch to other patch within a forest. So, habitat corridor development and gap reduction are the major issue to be addressed for effective animal connectivity. Safe animal connectivity is always preferable for both human society and biodiversity stated by Fernando et al. (2008). It may be managed by corridor demarcation and development which will allow safe wild species movement.

References

Ahmad F, Goparaju L, Qayum A (2018) Wild life habitat suitability and conservation hotspot mapping: Remote Sens GIS Based Decis Support Syst

Anon (2006–2007) State forest report. Forest department, Government of West Bengal. Directorate of forests office of the principal chief conservator of forests. Kolkata

Anderson JE (1991) A conceptual framework for evaluating and quantifying naturalness. Conserv Biol 5 (3):347–352

Austin M (2007) Species distribution models and ecological theory: a critical assessment and some possible new approaches. Ecol Model 200(1–2):1–19

Balfour DA, Bond WJ (1993) Factors limiting climber distribution and abundance in a southern African forest. J Ecol 93–100

Baltanás A (1992) On the use of some methods for the estimation of species richness. Oikos 484–492

Bera D, Chatterjee ND, Ghosh S, Dinda S, Bera S, Mandal M (2022) Assessment of forest cover loss and impacts on ecosystem services: coupling of remote sensing data and people's perception in the dry deciduous forest of West Bengal, India. J Cleaner Prod 356:131763

Chatterjee ND (2016) Man-Elephant conflict: a case study from forests in West Bengal. Springer, India

Chatterjee ND, Chatterjee S (2014) Changing Habitat and Elephant Migration from Dalma Wildlife Sanctuary, Jharkhand to Panchet forest division, Bankura, West Bengal: a biogeographical analysis. In: Climate change and biodiversity. Springer, pp 209–222

Chen JQ, Franklin JF (1990) Microclimatic pattern and basic biological responses at the clear-cut edges of old-growth Douglas-fir stands. Northwest Environ J 6 (2):424–425

Chowdhury S, Khalid MA, Roy M, Singh AK, Singh RR (1997) Management of elephant populations in West Bengal for mitigating man-elephant conflicts. Wildlife Institute of India, Dehradun

Das Gupta SP (1989) Forest eco-system in West Bengal. In: Budhadev C, Asok Maiti (ed) Forest and forest development in India. Inter-India Publications, D-17, Raja Garden Extn., New Delhi

Eltringham SK, Malpas RC (1993) The conservation status of Uganda's game and forest reserves in 1982 and 1983. Afr J Ecol 31(2):91–105

Elzinga CL, Salzer DW, Willoughby JW, Gibbs JP et al (2001) Monitoring plant and animal populations. Monit Plant Anim Populations

Farina A (2008) Principles and methods in landscape ecology: towards a science of the landscape, vol 3. Springer Science & Business Media

Fernando P, Wikramanayake ED, Janaka HK, Jayasinghe LKA, Gunawardena M, Kotagama SW, Weerakoon D, Pastorini J (2008) Ranging behavior of the Asian elephant in Sri Lanka. Mammalian Biology-Zeitschrift Für Säugetierkunde 73(1):2–13

Foster SL, Kharouba HM, Smith TW (2022) Testing the assumption of environmental equilibrium in an invasive plant species over a 130 year history. Ecography e12933

Forman RTT (1995) Land mosaic: the ecology of landscape and regions. Cambridge University Press, Cambridge, England

Forman RT (2012) Safe passages: highways, wildlife, and habitat connectivity. Island Press

Forman RT, Deblinger RD (2000) The ecological road-effect zone of a Massachusetts (USA) suburban highway. Conserv Biol 14(1):36–46

Gates DM (1980) Solar radiation. In: Biophysical ecology. Springer, New York, NY, pp 96–147

Goparaju L, Ahmad F, Sinha D (2018) Spatial analysis of wildlife habitat around Madihan forests of Mirzapur district, Uttar Pradesh in India, using geospatial technology

Gottschalk TK, Huettmann F, Ehlers M (2005) Thirty years of analysing and modelling avian habitat relationships using satellite imagery data: a review. Int J Remote Sens 26(12):2631–2656

Gustafson EJ, Parker GR (1992) Relationships between landcover proportion and indices of landscape spatial pattern. Landscape Ecol 7(2):101–110

Harper JL (1977) Population biology of plants. Popul Biol Plants

Hoare RE (1999) Determinants of human–elephant conflict in a land-use mosaic. J Appl Ecol 36 (5):689–700

Kulandaivel S (2010) A paradigm shift in the elephant depredation in South Bengal. Divisional Forest Officer, Bankura North Division, Personal Communication

Lembrechts JJ, Nijs I, Lenoir J (2019) Incorporating microclimate into species distribution models. Ecography 42(7):1267–1279

Li H, Reynolds JF (1995) On definition and quantification of heterogeneity. Oikos 280–284

Mandal M (2018) Forest range wise Asian Elephant's (Elephas Maximus) Habitat suitability assessment through food and water availability: a case study in Panchet Forest Division, Bankura, West Bengal

Mandal M, Chettarjee ND, Hazra J (2015) Elephant migration and colonization in Bankura district, West Bengal, India. Vidyasagar University. Indian J Geogr Environ 14:46–52

Mandal M, Chatterjee ND (2018) Quantification of habitat (forest) shape complexity through geo-spatial analysis: An ecological approach in Panchet forest division in Bankura, West Bengal. Asian J Environ Ecol 6:1–8

Mandal M, Chatterjee ND (2019) Forest core demarcation using geo-spatial techniques: a habitat management approach in Panchet forest division, Bankura, West Bengal India. Asian J Geogr Res 2(2):1–8

Mandal M, Chattarjee ND (2020a) Land use alteration strategy to improve forest landscape structural quality in Radhanagar forest range under Bankura district. Eurasian J For Sci 8(1):1–10

Mandal M, Chattarjee ND (2020b) Geo-statistical analysis to understand nature of forest patch shape complexity in panchet forest division under Bankura district West Bengal. Indian J Ecology 47(1):96–101

Mandal M, Chatterjee ND (2020c) Elephant's habitat suitability assessment through geo spatial quantification in Panchet forest division, West Bengal. Ecofeminism Clim Chang

Mandal M, Chatterjee ND (2021a) Geospatial approach-based delineation of elephant habitat suitability zones and its consequence in Mayurjharna Elephant Reserve, India. Environ Dev Sustain 23(12):17788–17809

Mandal M, Chattarjee ND (2021b) Estimation of forest ecosystem quality using GIS tool in Panchet forest division, West Bengal, India. In: Forest resources resilience and conflicts. Elsevier, pp 203–213

Mandal M, Chatterjee ND (2021c) Forest landscape and its ecological quality: a stepwise spatiotemporal evaluation through patch-matrix model in Jhargram District, West Bengal State India. Reg Sustain 2 (2):164–176

Mandal M, Chettarjee ND (2021d) Human-Elephant conflict in Joypur forest influence areas, West Bengal, India. Gajah 54:34–36

McGarigal K, Marks BJ (1995) FRAGSTATS: spatial pattern analysis program for quantifying landscape structure. General Technical Report PNW-GTR-351. Portland, OR: US Department of Agriculture, Forest Service, Pacific Northwest Research Station, vol 122 p, 351

McGarigal K, Tagil S, Cushman SA (2009) Surface metrics: an alternative to patch metrics for the quantification of landscape structure. Landscape Ecol 24(3):433–450

Nowakowski AJ, Frishkoff LO, Agha M, Todd BD, Scheffers BR (2018) Changing thermal landscapes: merging climate science and landscape ecology through thermal biology. Current Landscape Ecol Rep 3(4):57–72

Pranovi F, Franceschini G, Casale M, Zucchetta M, Torricelli P, Giovanardi O (2006) An ecological imbalance induced by a non-native species: the Manila clam in the Venice Lagoon. Biol Invasions 8(4):595–609

Shapiro A-MT, Harper SJ, Westervelt JD (2004) The fort hood avian simulation model-V: a spatially explicit population viability model for two endangered species. In: Landscape simulation modeling. Springer, pp 233–247

Singh AK (2006) Ecological investigation of human-elephant conflicts in South West Bengal. Saurashtra University

Singh AK, Singh RR, Chowdhury S (2002) Human-elephant conflicts in changed landscapes of south West Bengal India. Indian Forester 128(10):1119–1132

Sukumar R et al (2003) The living elephants: evolutionary ecology, behaviour, and conservation. Oxford University Press

Taylor PD, Fahrig L, Henein K, Merriam G (1993) Connectivity is a vital element of landscape structure. Oikos 571–573

Timilsina N, Ross MS, Heinen JT (2007) A community analysis of sal (Shorea robusta) forests in the western Terai of Nepal. For Ecol Manage 241(1–3):223–234

Turner MG (1989) Landscape ecology: the effect of pattern on process. Annu Rev Ecol Syst 20(1):171–197

Turner MG (1990) Spatial and temporal analysis of landscape patterns. Landscape Ecol 4(1):21–30

Turner MG, Gardner RH et al (1991) Quantitative methods in landscape ecology. Springer, New York NY

Watson JE, Whittaker RJ, Dawson TP (2004) Habitat structure and proximity to forest edge affect the abundance and distribution of forest-dependent birds in tropical coastal forests of southeastern Madagascar. Biol Cons 120(3):311–327

Wilcove DS, McLellan CH, Dobson AP (1986) Habitat fragmentation in the temperate zone. Conserv Biol 6:237–256

Woodward FI (2009) Biosphere—atmosphere interactions in landscapes. Princeton Guide Ecol 474

5 Species Specific Corridor Demarcation: Case of Asian Elephant

Abstract

Habitat corridor helps to understand functioning of energy, material as well as organism movement. It maintains landscape's resistance capacity. Ecological function may be positive or negative but both have influence on the movement of wildlife. Ecological behaviour of species in the landscape is a factor related to their movement pattern. This chapter tries to demarcate the potential corridor used by wild elephants that are migrating from Dalma Wildlife Sanctuary, Jharkhand State to South West Bengal State. It is based on habitat suitability model using major landscape attributes and Least Cost Path Analysis model. These models formulate the shortest route from sink point to source points through the highest suitable areas. The movement of elephants and their habitat functions track the path of least resistance across a landscape. This typical relationship gives rise to a distinctive nature of wildlife–habitat relationship which can be spatially modelled using GIS tools. The spatial model signifies the character of habitat corridor which elephant used to cover during migration.

5.1 Potential Elephant Corridor and Its Consequence

Species movement is one of the most important processes in ecology. Nature of movement depends on the combination of individual species behaviour and existing landscape function. Landscape alteration process is continuous, and it creates heterogeneity in habitat as well as in ecosystem stated by Fischer and Lindenmayer (2007). As a result ecosystem productivity devolves in a sectored way. Species search these productivities for their basic requirements and move to these productive places (López et al. 2020; Moniruzzaman et al. 2021; dos Santos et al. 2021). Therefore, ecosystem diversity in reality has a force to divert the flow of materials and energy in a comfortable direction. Species also maintain this rule for their movement. Space, time and disturbance regimes these three factors control the movement pattern in case of large size animal dispersal (Sukumar 2003; Shea et al. 2004; Bowler and Benton 2005; Shaw 2020). Another factor is fragility in ecosystem. In most of the cases, fragility comes from human civilization in the form of development,

M. Mandal and N. Das Chatterjee, *Geo-Spatial Analysis of Forest Landscape for Wildlife Management*, GIScience and Geo-environmental Modelling, https://doi.org/10.1007/978-3-031-33606-5_5

destruction and unethical performance of society stated by several environmental expert. In the same manner, natural event also responsible but somehow it is related to anthropogenic customs. Species avoids this fragility and makes a suitable way or corridor where disturbance is less for their movement from sink to source habitat (Mallegowda 2015; Pierik et al. 2016; Tong and Shi 2020). Therefore, safe and suitable corridor is preferred for selecting movement path by animal species.

South West Bengal is the part of extended home range of Dalma elephant. Elephant related ecological disturbance was not so frequent before 1980 addressed by Chowdhury et al. (1997), Singh (2006). Accidentally, individual occurrence was recorded in the foremost western portion of the state. After 1987, a herd of 40 elephants entered into the state following the regenerated forest patches (Sukumar 2003). Many researches pointed out the reason behind the surprising migration of elephant from Dalma Wildlife Sanctuary to the Jungles of South West Bengal. According to Lahiri-Choudhury (1996), Singh et al. (2002), Sukumar (2003), Chatterjee and Chatterjee (2014) and Chatterjee (2016), massive habitat destruction by development activities in adjoining Dalma Wildlife Sanctuary created imbalance in the existing ecosystem. As a result, elephants were forced to move towards the less disturbed areas in South West Bengal. This is surprising to note that the movement become regular in the study area (Mandal et al. 2015; Chatterjee and Mandal 2020). There is an increasing trend found with increasing number of elephant in a herd (Mandal and Chatterjee 2020c). Based on this situation, Government of West Bengal declared Mayurjharna Elephant Reserve on 24 October 2002 to conserve or to protect elephant movement and colonization. Recent, published articles and reports stressed on the fact that elephants now become colonised in this region due to suitable landscape productivity.

Long ranging wildlife movement route is very sensible for biodiversity as well as human civilization (van Toor et al. 2018). The range is different in different wildlife domain. Such as elephant (*Loxodonta africana*) in Africa covers longer distance than Asian elephant (*Elephas maximus*) for seasonal movement (Baskaran et al. 2010). The nature of movement and corridor is also different in these two countries. But elephants of both the countries select their corridor based on landscape diversity (Baskaran et al. 2018). They also use same corridor at the time of return from source to sink region which are free form disturbance (Chatterjee 2016). So many ecological issues arised from the elephant movement corridor. One of the most important issue is Human–elephant conflict (HEC). Such phenomena in South West Bengal become burning issue for both biodiversity and human society (Kumara et al. 2017; Mandal and Chatterjee 2020c). It has been found that huge **amounts** of crops and assets are damaged through **out** the elephant corridor in this region. Human death is also an incident in this process. Not only that, elephant death or injury by human society is also a common phenomena. Such kind of circumstances will be controlled or managed if we have the idea about the elephant corridors throughout the region. Therefore, corridor identification is very essential to save the society as well as the wildlife.

A comprehensive empirical investigation is the most scientific way to demarcate elephant corridor in this region. But this process is very time consuming and expensive too. For that reason, remote sensing data are used in worldwide to predict the possible corridors. GIS and remote sensing become an essential platform for investigating of any kind of landscape ecological function (Turner et al. 2004; Miller and Rogan 2007; Yu et al. 2019). Satellite images and various spatial analysis programmes were used to generate prediction models (Roy and Tomar 2000; Gülçin 2020). Such kind of predictive models have been used in ecology for stimulating different problems and provide potential possibilities (Boyd and Foody 2011). The present study applied such kind of technique to demarcate potential elephant corridor in the study area.

5.1.1 Assessment Methods

Wildlife activities do not fallow any administrative boundary. In south-east India, elephant habitat mostly covers the states of Jharkhand, Orrisa and West Bengal. The study has been carried out in some parts of these states. Elephant dispersal and migration is seen in this area (Fig. 5.1). The affected area lies between 86° 0′ 25.73″ E to 87° 33′ 15.23″ E and 22° 19′ 0.30″ N to 23° 59′ 57.74″ N. It covers 21,415.19 km^2 areas. The most affected state is West Bengal according to Santra et al. (2007), Kulandaivel (2010). Four districts Puruliya, Bankura, Paschim Medinipur and Jhargram in South West Bengal experienced elephant movement related issues throughout the year. The adjoining areas of neighbouring state, Jharkhand is considered for this study because of elephant sink region is located in this state.

This region encompasses the eastern part of Chota Nagpur Plateau. Dalma hill in Jharkhand State is the highest altitude point in the study region, and it gradually downs towards the east. The entire study area falls under the catchment areas of four major rivers Subarnarekha, Kangsabati, Silabati and Dwarakeswar. The uplands of inter-fluvial zone of these rivers covered with lateritic soil and composed of tropical dry deciduous forest ecosystem mainly dominated by sal (*Shorea robusta*) forest. The forest is usually highly fragmented (Mandal and Chatterjee 2018) and patchy in nature (Mandal and Chatterjee 2019). Agricultural land is the dominant landuse in this landscape mosaic.

5.1.2 Data Source and Preparation

Landuse is the key attribute for understanding the elephant movement behaviour (Desai and Hedges 2010). Elephant habitat preference led to select suitable corridor for movement within the forage ground. Therefore, habitat suitability model is very important to know the movement

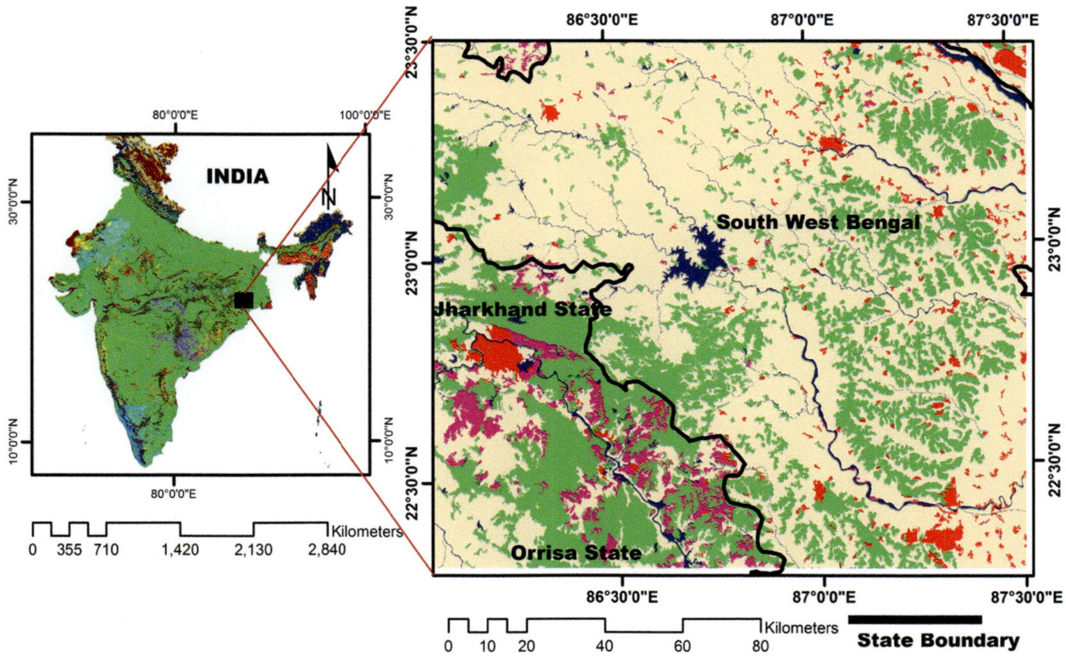

Fig. 5.1 Location of the study area: South West Bengal, Jharkhand and Orrisa States in India

characters of elephant. To carry out this model landuse land cover (LULC) classification has been made with the help of ENVI 5.2, ArcGIS 10.3 and ERDAS 9.3 version software using Landsat 8 (OLI/TIRS) satellite image collected from Earth Explorer, United States Geological Survey (USGS) site (http://glovis.usgs.gov/). Techniques like supervised classification, Spectral Angle Mapping (SAM) and Support Vector Machine (SVM) algorithm are implemented in the Landsat image of 2019 for the LULC classification (Pal and Mather 2004, 2005; Bouaziz et al. 2017). In this study, based on the spatial coverage of the study areas, about 12 points of each class were selected for the LULC classification. The five LULC classes have been extracted finally on the basis of research objectives (Fig. 5.2). After LULC classification, the entire study area was divided into 864 grids by using Fishnet tool in ArcGIS 10.3 v software. Each grid covers 25 km^2 areas. This grid framework map then overlayed on the LULC map. Individual landuse class density is prepared with the help of this information. These are very essential input for habitat suitability modelling. Another objective of the study is to identify the sink and source region. Ecologically, sink region is lacking in native species, and source region offers better opportunities for elephants. Within the spatial coverage of the study area, elephant sink region is well documented by Mishra and Nayak (2010), Chatterjee and Chatterjee (2014), Guha and Guha (2014). Sink region is in Dalma Wildlife Sanctuary in Jharkhand State from where the elephant moves to different source region (Fig. 5.2). To identify the source region in the districts of South West Bengal, reliable documents and expert opinions were collected. According to many scholars Singh (2006), Kulandaivel (2010), Chatterjee (2016), Mandal

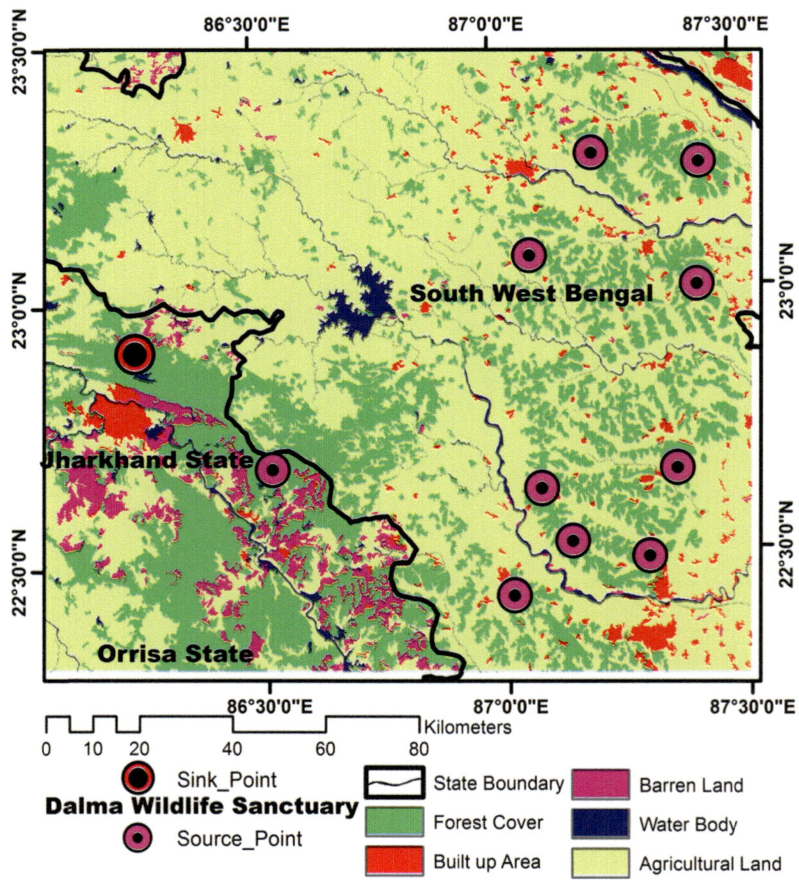

Fig. 5.2 LULC map and location point of sink and source region for elephant movement in the study area

and Chatterjee (2020c) they addressed some particular forest areas in these districts where elephant herds usually entered and come back from these forests without extending the movement further. The study considered 10 such forest habitat patches at the source point and also pointed out those places in LULC classification map (Fig. 5.2).

5.1.3 Habitat Suitability Model

Elephant continue their movement in search of suitable habitat from sink region to source region. This research aims is to demarcate the movement corridor of elephant in this study area, and habitat suitability mapping is very essential for that. Suitable habitat offers the function where elephant feel safety from disturbance and get basic requirements easily. The ecological disturbances, safety and available food and water are depending on existing landscape setup (Naveh and Lieberman 2013). In heterogeneous, human-intruded landscapes, distribution, food availability, and calm environment are hampered. So, it is clear that LULC pattern directly control the habitat suitability. Each individual landuse classes like forest, agriculture, barren land, built-up area, water body, road network line, etc., may act as a barrier to habitat suitability where other geospatial attributes like slope, relative height, climate, etc., are in comfortable in condition. Elephant when move from one suitable habitat to next habitat they usually chose the path based on these criteria. suitability (Pandit and Chanda 2019; Mandal and Chattarjee 2020d). On the basis of these selected landuse, the present study carried out a habitat suitability model by using reliable LULC inputs in ArcGIS 10.3 v software interface. Individual landuse density distribution is the essential input to run this model.

5.1.4 Model Variables

Based on previous research outcomes (Singh et al. 2002; Baskaran et al. 2013; Bastille-Rousseau et al. 2018) and interviews with elephant experts, the study selected major six landuse classes for habitat suitability mapping. These are forest cover, agriculture land, barren land, built-up area, water body, road network line (Rail line and motor vehicle road). The entire study area divided in grid framework (Fig. 5.3b). Grid wise individual class amount has been counted by intersect and merge tool in ArcGIS 10.3 v software. On the basis of

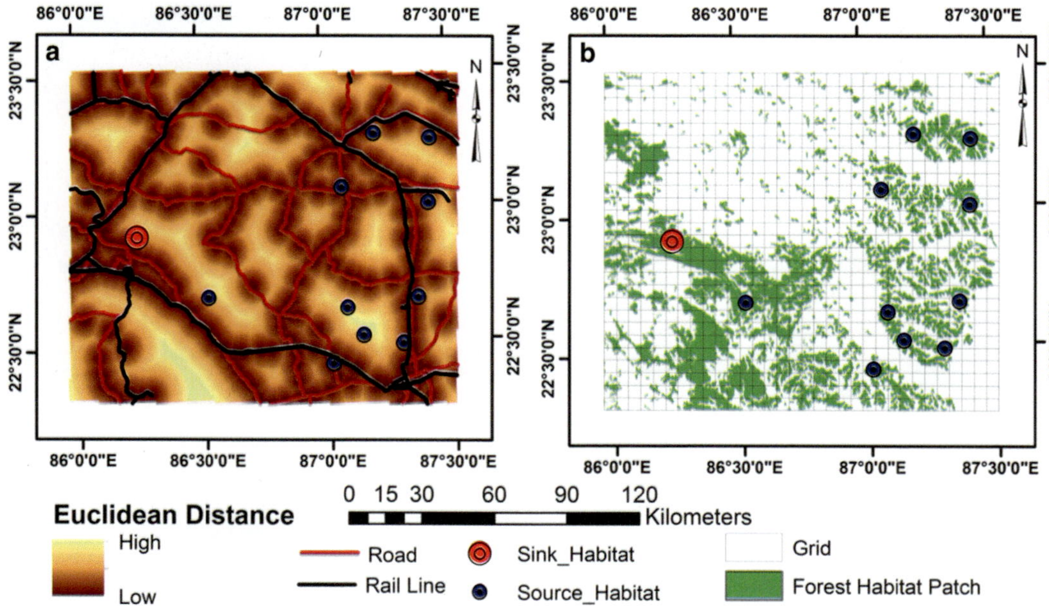

Fig. 5.3 Euclidean distance of road network line (**a**) and grid framework map of the study area (**b**)

these grid value, density maps were prepared (Fig. 5.4) using the kriging method in interpolation tool in the same software except road network line. Major and minor roads under the study area were digitized from open source Google image 2019. A Euclidean distance map was prepared (Fig. 5.3a) using these line features in same software.

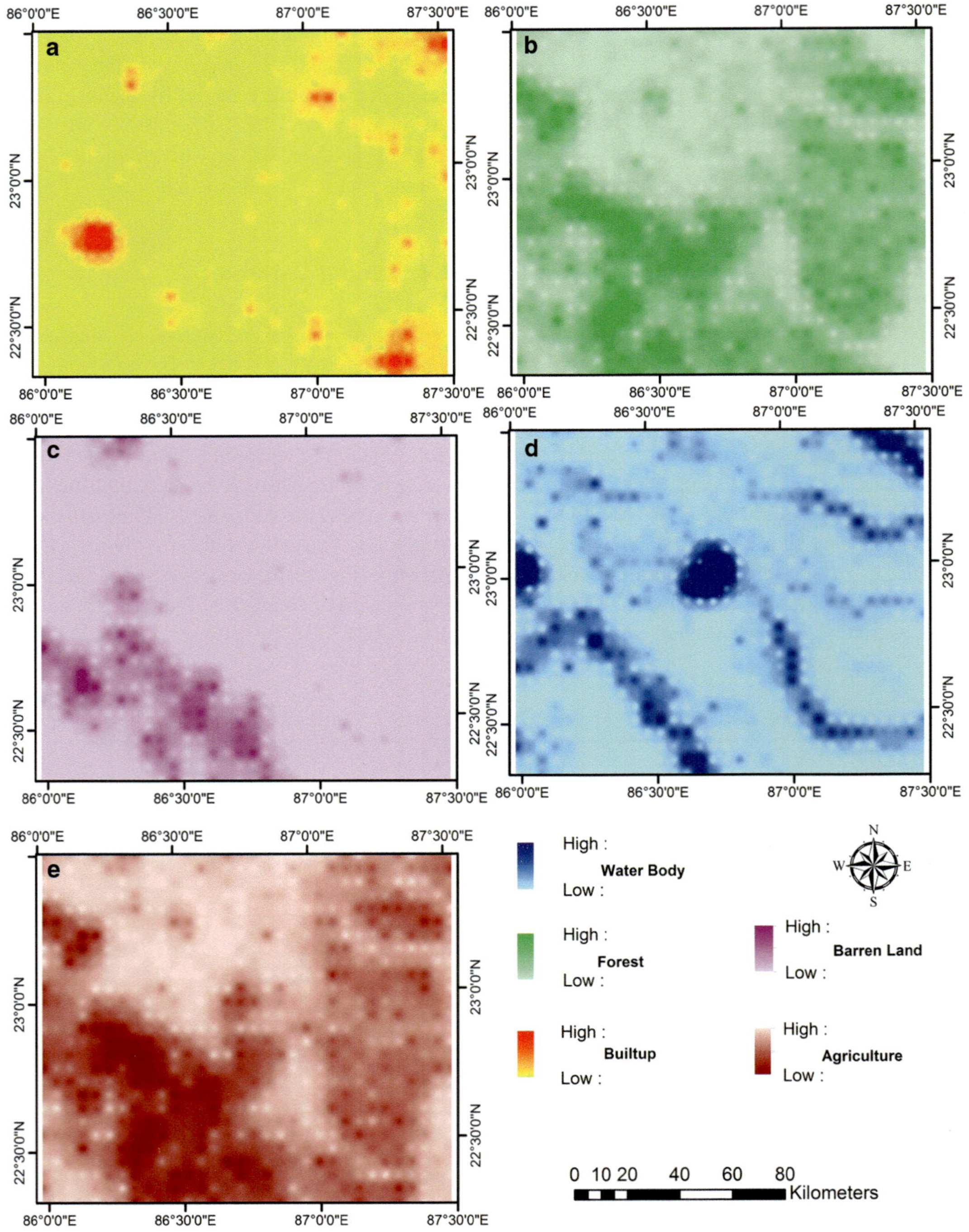

Fig. 5.4 Density maps of built-up area (**a**), forest cover (**b**), barren land (**c**), water body (**d**) and agricultural land (**e**) under the study area

5.1.5 Variable Resistance Weights

Every landscape has specific ecological resistance and resilience that have positive or negative effects on individual species' habitat quality (Harrison and Bruna 1999; Altermatt and Holyoak 2012; Baguette et al. 2013). This resistance capacity responds diversely for different species habitat. At large landscape scales, different sites may exhibit varied levels of resistance and resilience due to spatial variability in environmental conditions, biological interactions, and disturbance legacies. There is a nonlinear relationship between the amount of habitat and the probability of species persistence. Resource availability, patch size, habitat connectivity, and disturbances are seen as major players defining the composition of natural communities. It may be positive or negative depends on individual species' biological behaviour with the landscape. Another component is the scale or weight of resistance of different landuse on species specific habitat quality. In case of elephant habitat quality some landuse has direct impact and has strong weight like forest cover and water body (Kija et al. 2020). There are some landuse also like barren land, crop land, built-up land have an impact but less than forest or water body. Another landuse like transport line and canal have a role in elephant habitat quality but it reacts negatively. So, in case of elephant habitat suitability assessment, the selected LULC weight measurement is very sential. The present study takes several expert opinion and the similar methods of Roy et al. (2010) to measure the weight in Orrisa State in India for ecological corridor mapping, and Li et al. (2010) for identify a giant panda dispersal corridor in Wolong Nature Reserve in China, and Mandal and Chatterjee (2020c) for elephant habitat suitability assessment in Panchet forest division in South West Bengal. Finally, in order to determine the weights of selected LULC, the present study performed an Analytic Hierarchy Process (Satty 1980, 1999) model using selected parameters. This is the popular multi-criteria decision making method for assessing the relative weights of different factors in a model. The selected six LULC resistance weights and functional weights are listed in Table 5.1.

5.1.6 Habitat Suitability Assessment in Arc Model

The present study builds a habitat suitability model using the ModelBuilder within ArcGIS 10.3 v software. Individual vector data (LULC) are converted in raster format and reclassified for assigning the weight according to Table 5.1. Afterwards, all selected reclassified and weight assigned raster layers of selected variables were added up using the Weighted Sum Tools in Arc Toolbox to create a single habitat suitability

Table 5.1 Variable weights and function for assessing habitat suitability index

Selected LULC	Resistance weight (%)	Density scale (equal interval)	Functional weight	
Forest cover	30	High Low	5 1	Positive (+)
Water body	19	High Low	5 1	Positive (+)
Agricultural land	17	High Low	5 1	Positive (+)
Built-up area	12	High Low	1 5	Negative (−)
Barren land	10	High Low	5 1	Positive (+)
Road network line	12	High Low	1 5	Negative (−)
	100			

map. The final map is reclassified in four equal classes.

5.1.7 Least Cost Path (LCP) Analysis

LCP analysis depend on graph theory (Bunn et al. 2000). It is continued by a raster-based algorithm that weights the minimal cost distance between a source point and destination (target) point. So the source and target cell are special input in this model. We have already described in upper section about these. The algorithm propagates through the raster cell. Raster cell resistance is the basic concept to run this algorithm. Surface resistance cost is weighted and combined from variables facilitating (positive) or hindering (negative) effects on considerable movement process. In case of wildlife movement, the habitat (LULC) resistance value linked with species habitat relationship. It calculates a minimum accumulated cost surface from over the entire landscape. The output raster map where every raster cell is a weight value, and it represents the lowest possible cost.

5.2 Assessment Results

The total research work has two broad segments. One is elephant habitat suitability mapping. Another is potential corridor demarcation through the habitat suitability class and its surface resistance capacity by LCP analysis.

5.2.1 Habitat Suitability Classes

The reclassified elephant suitability map of the study area categorized in four equal classes. Hierarchically, these are high suitable class cover only 1099.72 km^2 (5.42%) areas, moderate suitable class covers 5549.03 km^2 (27.39%) areas, low suitable class covers 10,261.83 km^2 (50.66%) areas and very low suitable class cover 3344.23 km^2 (16.51%) areas under the study area. It has been found from this classified map, most of highly suitable areas underlying over the dense forest cover landuse. The marking elephant migration source or destination points are mostly located in moderate suitable class (Fig. 5.5). But elephant habitat sink point, i.e. Dalma Wildlife Sanctuary is under highly suitable area. Insufficient food supply, frequent mining pit, enormous transport development activities force them to move towards less suitable areas stated by Chatterjee and Chatterjee (2014). Elephant don't move towards south west portion of the study areas where habitat condition is better because of food unavailability. Becoming a large territorial animal, elephant requires huge amount of food and water for their biological assimilation (Sukumar 1990, 2003). For that reason, elephant moves towards suitable areas where food and water become available throughout the year (Fernando et al. 2008). In north and north-east portion of the study area, agricultural crops are practised enormously throughout the year and forest quality of forest was improved by social forestry (Chatterjee and Mandal 2020). As a result of this, elephant likes to migrate in these areas where habitat is moderately suitable but foods are easily available and habitat disturbance is very less.

5.2.2 Potential Elephant Corridors

This is the main objective of this study. For this purpose, elephant habitat suitability index is prepared. Using this habitat suitability resistance LCP model, we finalized the elephant potential corridors between habitat sink point and source points. A total of 599.70 km potential corridor route is demarcated under the study area (Fig. 5.6). It is observed that 248.51 km elephant corridor passes through the moderate suitable zone which is the highest corridor length out of total. 235.21 km potential corridor passes through less suitable area and 30.75 km in very less suitable area. Only 85.3 km corridor crosses through the highly suitable areas. Human activities are frequent in moderate and less suitable area. This result shows that HEC may be more frequent in this region because of exposure of elephant from highly suitable areas to less suitable areas over and over again. Six nodal points

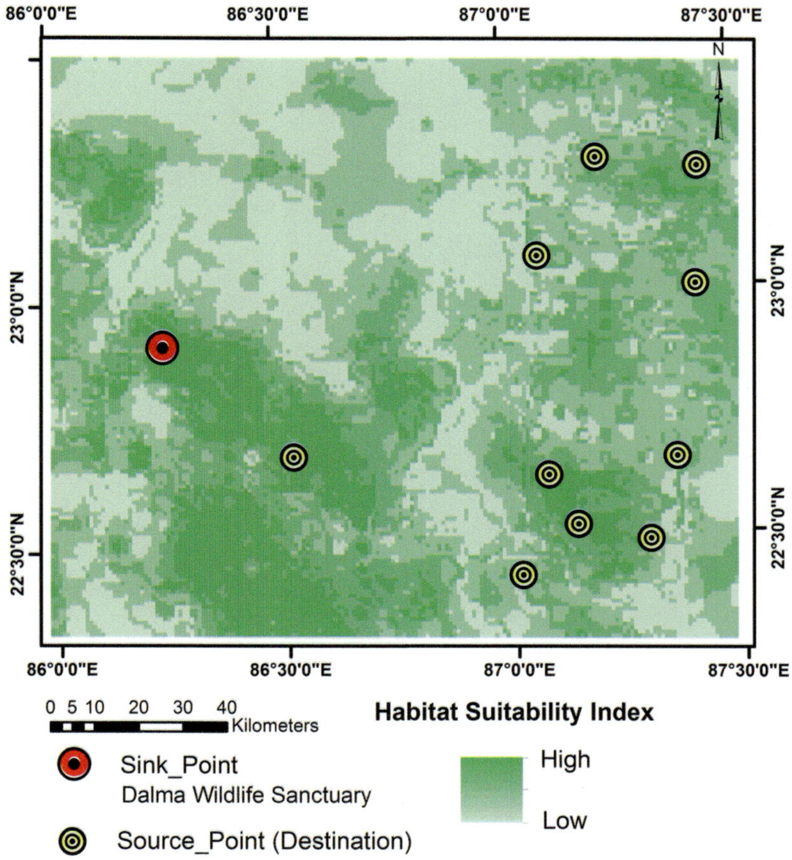

Fig. 5.5 Habitat suitability index map of the study area with elephants habitat source and sink points

are present between this potential migration route. From these nodal points, elephant may be diverted their route to arrive to the different source or target habitat points (Fig. 5.6). Five nodal habitat points out off six are located in Mayurjharna Elephant Reserve and only one nodal habitat point is located in between Bankadaha and Bishnupur-I Forest Beat under Panchet Forest Division in Bankura distract. From Mayurjharna Elephant Reserve, elephants may move to Jhargram, the destination habitat point through the States of Jharkhand and Orrisa. Other potential corridors from sink to source points are described in details in Table 5.2.

5.2.3 Potential Elephant Corridor Zone

LCP analysis predicted the linear cost distance corridor from sink habitat to different source or target habitat points. These are the probable way to move or to migrate from sink to source points. The present study further presented elephant corridor potential zone in four classes. These are highly potential area covers 650.43 km^2, moderate potential area covers 1333.89 km^2, low potential area covers 1489.80 km^2 and very low potential area covers 1311.92 km^2 along with the straight corridor line (Fig. 5.7). In highly potential corridor zones, elephant herd could take rest and shelter where habitat qualities are better, and disturbance activities are minimal. Highly potential corridor zones are functioning as species colonizing centre stated by Roy et al. (2010) in ecological corridor demarcation in Orissa State. Li et al. (2010) also stated similar approach in Giant Panda dispersal behaviour in Wolong Nature Reserve in China and Bennett and Mulongoy (2006) found similar results for elephant colonization in Africa. Elephant herd from Dalma Wildlife Sanctuary are colonized in these places

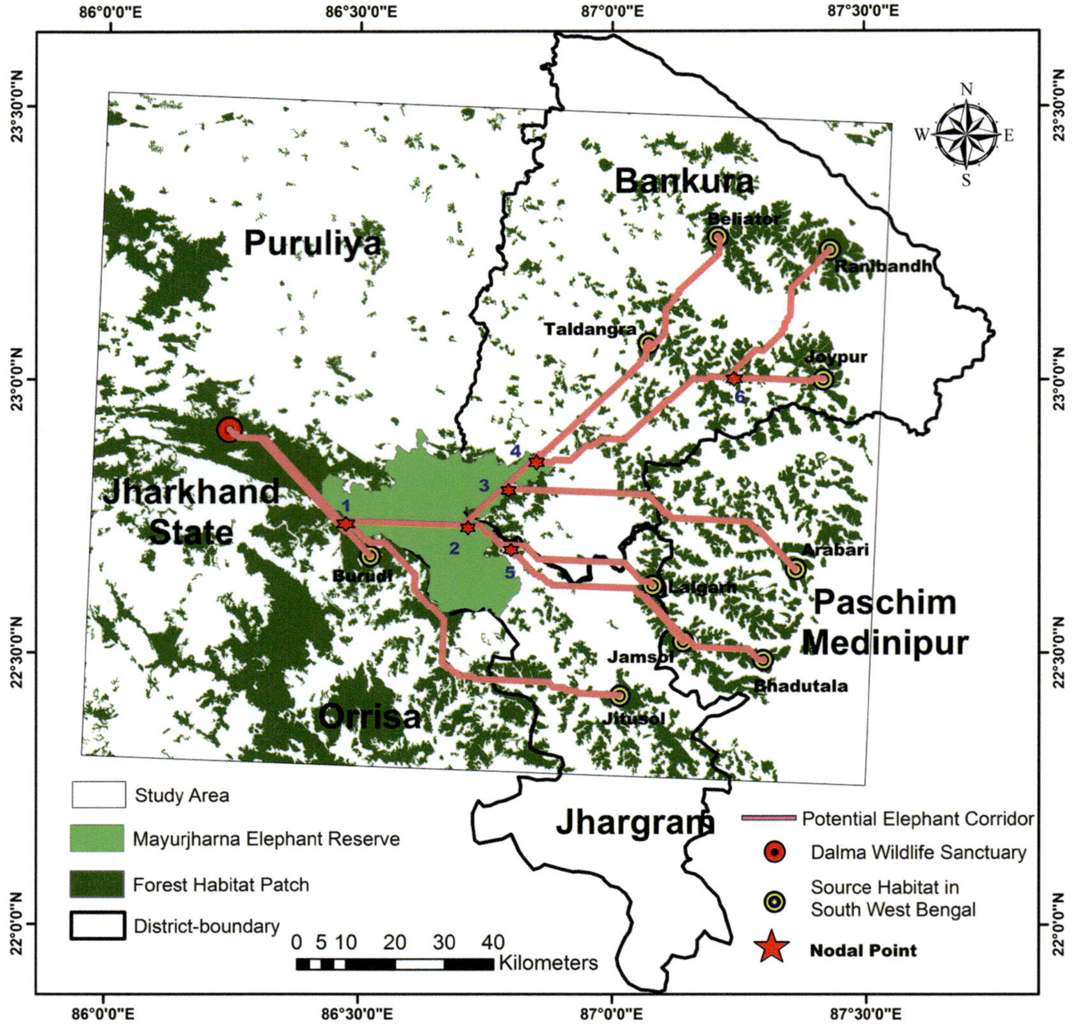

Fig. 5.6 Potential elephant movement corridors from Dalma Wildlife Sanctuary to selected different destination points under the study area

and move towards different target place in small groups. So biogeographically, these areas have more attention for managing the elephant migration issues and protect biodiversity as a whole. In the study area, five highly potential corridor zones are identified. Three highly potential corridor zones (Z-1, 2 and 3) are located under Mayurjharna Elephant Reserve. Jhitkar Jungle near Lalgarh is another highly potential zone (Z-4) under Medinipur Forest Division and another one (Z-5) is located near Peardoba jungle under Bankadaha Forest Range (Plate 5.1 and Fig. 5.7).

The LULC distributions along with different potential corridor zones represent an interesting issue. Individual landuse class plays a significant role for characterizing the elephant movement pattern (Sukumar 2003). It has been measured that percentage of forest land cover gradually increases from very low to high potential corridor zone (Fig. 5.8). The same trend is found for water body. These two landscape elements improve the elephant habitat quality as well as it helps for selecting favourable corridor path. Percentage of Agricultural land and built-up area are following the negative trend. It means the percentage is declined with increasing corridor potentiality. Higher percentage of positively functioned LULC especially forest cover and

Table 5.2 Potential corridor distance from Dalma Wildlife Sanctuary to different source or destination points through nodal points

Sink point	Nodal points location		Distance in km from sink point to nodal points	Source or destination points	Distance in km from nodal point to destination
Dalma Wildlife Sanctuary, Jharkhand	Mayurjharna Elephant Reserve	• Nodal point-1	31.83	Burudi	9.20
				Jitusol	76.01
		• Nodal point-2	55.23		
		• Nodal point-3	64.41	Aabari	129.73
		• Nodal point-4	74.66	Taldangra	32.88
				Beliatore	61.11
		• Nodal point-5	65.43	Lalgarh	32.44
				Bhadutala	31.89
				Jamsol	64.71
	Bankadaha Forest range	• Nodal point-6	120.74	Ranibandh	34.78
				Joypur	19.45

Plate 5.1 Forest habitat characters in highly potential zone in **a** Peardoba under Bankadaha forest range, **b** Joypur forest range in Bankura district, **c** Jhitka forest near Lalgarh and **d** forest under Mayurjharna Elephant Reserve

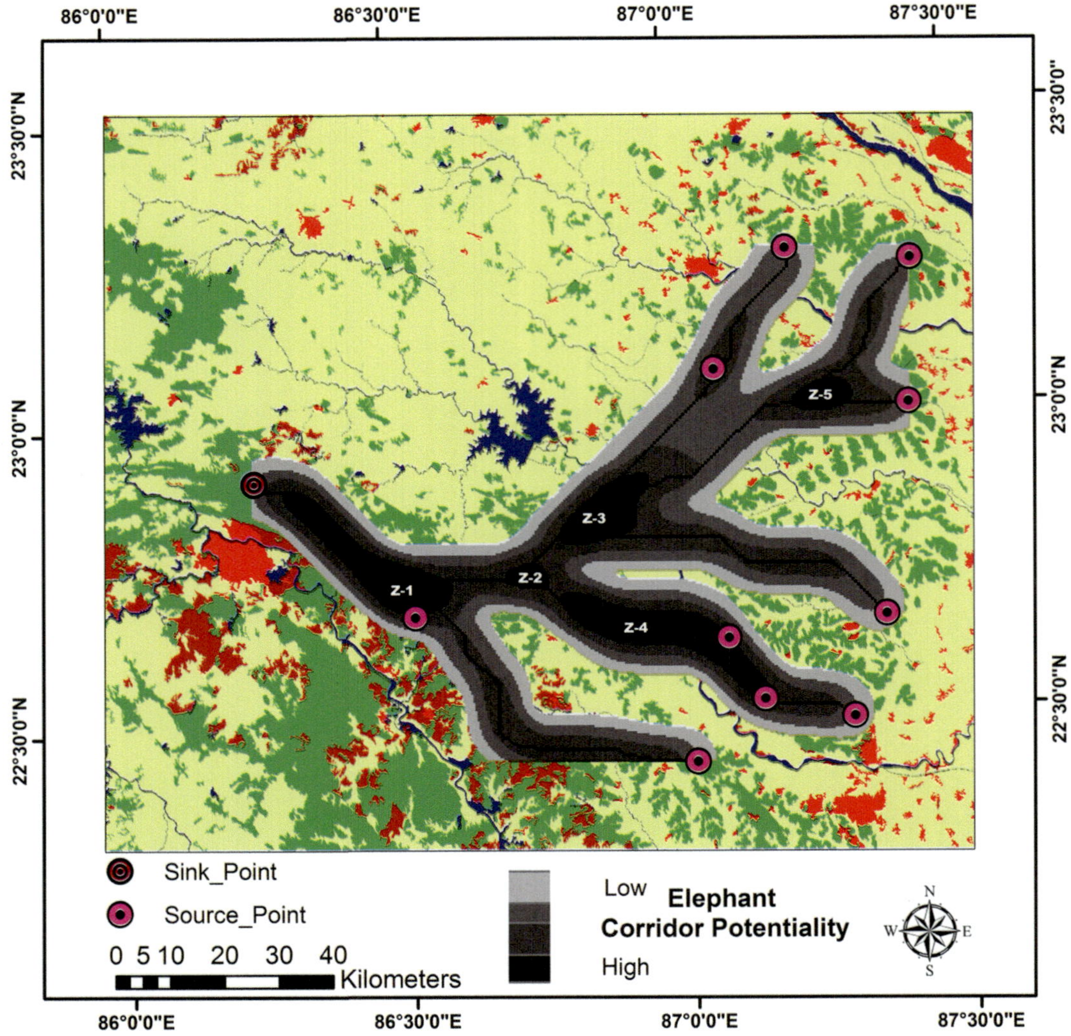

Fig. 5.7 Classified potential elephant corridor zones in the study area

water body and very less percentage of disturbing landuse like built-up area in highly potential corridor zone make it more favourable for elephant to stay and to move securely.

5.2.4 Potential Elephant Corridor Characters

Corridor landscape is another aspect for predicting its characters. Not percentage but the amount of landuse along the corridor influences the corridor character. In the study area, 599.70 km corridor was demarcated. 302 km corridor length out of total passes through the forest cover which is the longest corridor. The second longest corridor 254.40 km passes through the agricultural land (Fig. 5.9) which covers 42.42% of the total corridor length. So elephant when moves to sink point to source points they frequently uses the agricultural land as their suitable movement path. This movement character will make huge crop damage in the study area (Plate 5.2).

Along with the demarcated elephant corridor, fourteen major road crossing points were identified. These transition points will create an imbalance between human and wildlife

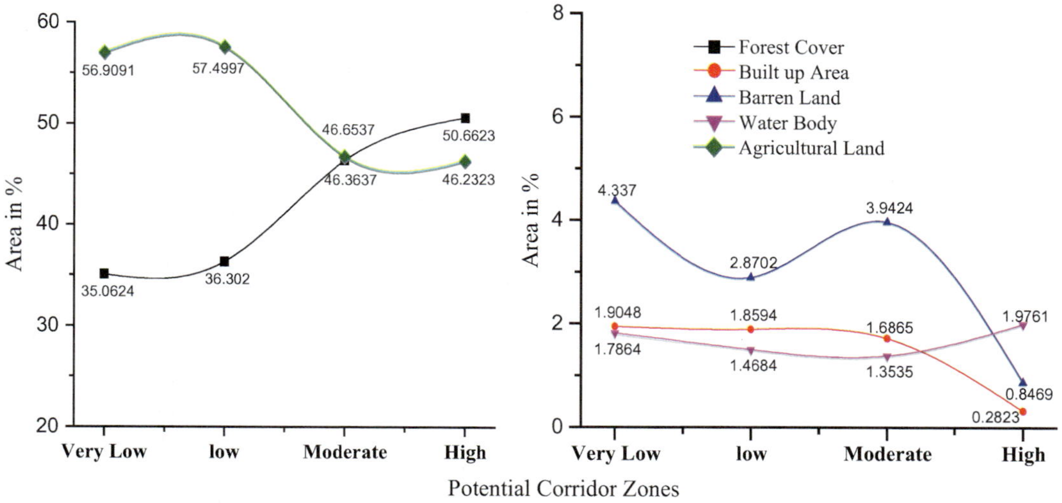

Fig. 5.8 LULC percentage area distribution in different corridor potential zones

Plate 5.2 Elephant herd raiding crops in their movement path near Salboni, Medinipur Forest Division. *Source* Rakesh Singha Dev

Fig. 5.9 Distribution of corridor length against amount of used LULC

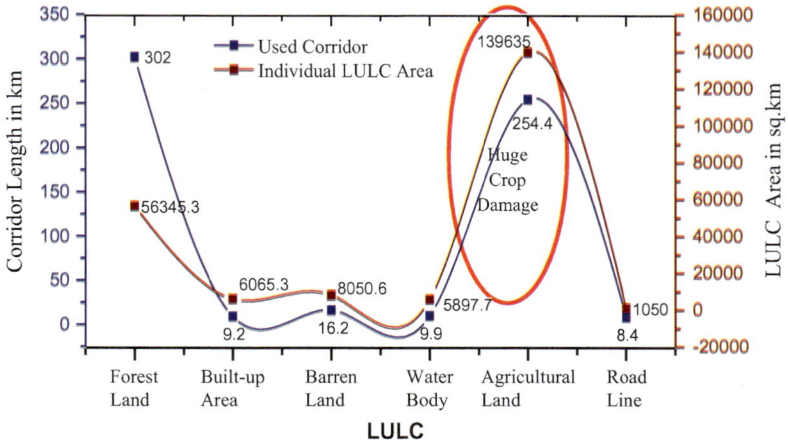

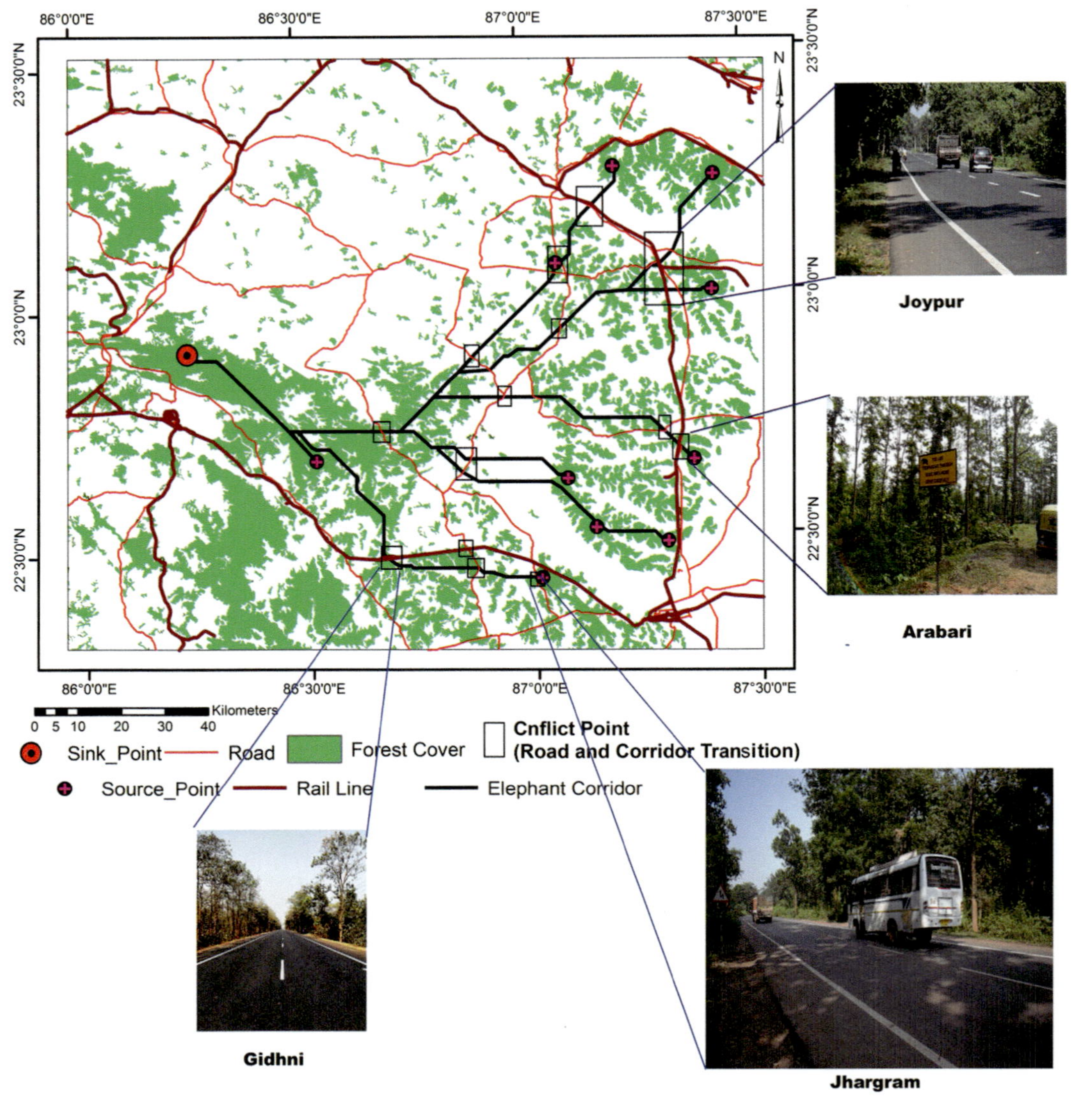

Fig. 5.10 Human–elephant corridor transition points in the study area

behaviour. Human corridors create several obstacles for elephant movement stated by several scholars (Fernando et al. 2008; Baskaran et al. 2013; Chatterjee and Mandal 2020). In Joypur, Jhargram and Arabari forest area NH-6, NH-60, South-Eastern rail way, Kharagapur to Adra rail line, and many State Highways overlapped with elephant potential corridors. So the roads crossing forage ground of elephant may hamper free movement in these transition points (Fig. 5.10; Plate 5.3).

5.3 Overall Outcomes

This is a GIS-based model that predicts the potentiality of occurrences. The occurrence of elephant movement through this potential corridor route is observed. The effectiveness of Predicted corridors depends on elephant movement behaviour and background landscape resistance. So changes in resistance variables ultimately control the animal behaviour and total process. In

Plate 5.3 Human corridor across the Arabari forest sometimes create noise that leads haphazard movement of wild animal

South West Bengal, elephant migration related issues are increased from the past (Singh 2006; Chakraborty and Mondal 2013; Chatterjee and Mandal 2020). Simultaneously, forest cover is also generated without maintaining any structural ecological quality (Mandal and Chattarjee 2020b). This led to forest fragmentation (McGarigal and Marks 1994; Forman 1995; Farina 2008; Taubert et al. 2018). Fragmented forest promotes irregular movement of wildlife from source to target habitat patch. In this patchy landscape, wildlife movement process and pattern increase the probability of human-wildlife conflict. To manage this situation elephant corridor demarcation is necessary. The present model proposes such map that presents potential corridor for South West Bengal and its adjoining areas. This study also appraised potential corridor zones that reveals the intensity of elephant movement through this corridor. Habitat development strategy through this corridor may be a suitable approach that may be useful for managing arbitrary movement of elephant in this region. This work may also be useful to find out the gap along the potential corridor of elephant. Habitat gap demarcation and management are ecologically essential for improving habitat quality. Monitoring of elephant movement and proper announcement may alert the adjacent villages will save many lives and assets along these corridors.

References

Altermatt F, Holyoak M (2012) Spatial clustering of habitat structure effects patterns of community composition and diversity. Ecology 93(5):1125–1133

Baguette M, Blanchet S, Legrand D, Stevens VM, Turlure C (2013) Individual dispersal, landscape connectivity and ecological networks. Biol Rev 88 (2):310–326

Baskaran N, Kanakasabai R, Desai AA (2018) Influence of ranging and hierarchy on the habitat use pattern by Asian elephant (Elephas maximus) in the tropical forests of Southern India. In: Indian hotspots. Springer, Singapore, pp 345–358

Baskaran N, Kannan G, Anbarasan U, Thapa A, Sukumar R (2013) A landscape-level assessment of Asian elephant habitat, its population and elephant–human conflict in the Anamalai hill ranges of southern Western Ghats, India. Mamm Biol 78(6):470–481

Baskaran N, Udhayan A, Desai AA (2010) Status of the Asian elephant population in Mudumalai Wildlife Sanctuary, southern India. Gajah 32:6–13

Bastille-Rousseau G, Douglas-Hamilton I, Blake S, Northrup JM, Wittemyer G (2018) Applying network theory to animal movements to identify properties of landscape space use. Ecol Appl 28(3):854–864

Bennett G, Mulongoy KJ (2006) Review of experience with ecological networks, corridors and buffer zones. In: Secretariat of the Convention on Biological Diversity, Montreal, Technical Series, vol 23, p 100

Bouaziz M, Eisold S, Guermazi E (2017) Semiautomatic approach for land cover classification: a remote sensing study for arid climate in southeastern Tunisia. Euro Mediterr J Environ Integr 2(1):24

Bowler DE, Benton TG (2005) Causes and consequences of animal dispersal strategies: relating individual behaviour to spatial dynamics. Biol Rev 80(2):205–225

Boyd DS, Foody GM (2011) An overview of recent remote sensing and GIS based research in ecological informatics. Eco Inform 6(1):25–36

Bunn AG, Urban DL, Keitt TH (2000) Landscape connectivity: a conservation application of graph theory. J Environ Manage 59(4):265–278

Chakraborty K, Mondal J (2013) Perceptions and patterns of human–elephant conflict at Barjora block of Bankura district in West Bengal, India: insights for mitigation and management. Environ Dev Sustain 15 (2):547–565

Chatterjee D, Mandal M (2020) Human-elephant conflict in Panchet Forest Division, Bankura, West Bengal, India. Gajah 51:10–15

Chatterjee ND (2016) Man-elephant conflict: a case study from forests in West Bengal. Springer International Publishing, India

Chatterjee ND, Chatterjee S (2014) Changing habitat and elephant migration from Dalma Wildlife Sanctuary, Jharkhand to Panchet Forest Division, Bankura, West

Bengal: a biogeographical analysis. In: Climate change and biodiversity. Springer, Tokyo, pp 209–222
Chowdhury S, Khalid MA, Roy M, Singh AK, Singh RR (1997) Management of elephant populations in West Bengal for mitigating man-elephant conflicts. Wildlife Institute of India, Dehradun
Desai AA, Hedges S (2010) Notes from the Co-chairs IUCN/SSC Asian Elephant Specialist Group. Gajah 33:3–6
dos Santos JS, Dodonov P, Oshima JEF, Martello F, de Jesus AS, Ferreira ME, Collevatti RG (2021) Landscape ecology in the Anthropocene: an overview for integrating agroecosystems and biodiversity conservation. Perspect Ecol Conserv
Farina A (2008) Principles and methods in landscape ecology: towards a science of the landscape, vol 3. Springer Science & Business Media
Fernando P, Kumar MA, Williams AC, Wikramanayake E, Aziz T, Singh SM (2008) Review of human-elephant conflict mitigation measures practiced in South Asia. WWF, Gland, Switzerland
Fischer J, Lindenmayer DB (2007) Landscape modification and habitat fragmentation: a synthesis. Glob Ecol Biogeogr 16(3):265–280
Forman RTT (1995) Land mosaic: the ecology of landscape and regions. Cambridge University Press, Cambridge, England
Guha M, Guha A (2014) Migration ethology of Dalma Herd and analysis of the rationale behind. The Beats Nat Sci 2(1)
Gülçin D (2020) The analysis of current approaches for evaluating landscape ecology. Contemp Stud Sci 195
Harrison S, Bruna E (1999) Habitat fragmentation and large-scale conservation: what do we know for sure? Ecography 22(3):225–232
Kija HK, Ogutu JO, Mangewa LJ, Bukombe J, Verones F, Graae B, Nzunda EF (2020) Spatio-temporal changes in wildlife habitat quality in the Greater Serengeti ecosystem. Sustainability 12 (6):2440
Kulandaivel S (2010) A paradigm shift in the elephant depredation in South Bengal. Divisional Forest Officer, Bankura North Division, Personal Communication
Kumara HN, Azeez PA, Singh A, Pal A, Nadu T (2017) Ecology of elephant (Elephas maximus) in South-West Bengal including population dynamics, migratory pattern, feeding habits and human-elephant conflict. In: Annual Report (April 2016–March 2017). Sálim Ali Centre for Ornithology and Natural History, Coimbatore, Tamil Nadu
Lahiri-Choudhury DK (1996) Hathi O Banjangaler Katha ("Tales of Elephants and Forests", in Bengali). Kolkata, Pratikshan
Li H, Li D, Li T, Qiao Q, Yang J, Zhang H (2010) Application of least-cost path model to identify a giant panda dispersal corridor network after the Wenchuan earthquake—Case study of Wolong Nature Reserve in China. Ecol Model 221(6):944–952
López S, López-Sandoval MF, Gerique A, Salazar J (2020) Landscape change in Southern Ecuador: An indicator-based and multi-temporal evaluation of land use and land cover in a mixed-use protected area. Ecol Ind 115:106357
Mallegowda P (2015) Functionality of wildlife corridors in the fragmented landscape of the Western Ghats, India: implications for conservation and management (Doctoral dissertation, Manipal University, Ashoka Trust for Research in Ecology and the Environment (ATREE))
Mandal M, Chatterjee ND (2018) Quantification of habitat (forest) shape complexity through geo-spatial analysis: an ecological approach in Panchet forest division in Bankura, West Bengal. Asian J Environ Ecol 6:1–8
Mandal M, Chatterjee ND (2019) Forest Core demarcation using geo-spatial techniques: a habitat management approach in Panchet Forest Division, Bankura, West Bengal, India. Asian J Geogr Res 1–8
Mandal M, Chattarjee ND (2020b) Land use alteration strategy to improve forest landscape structural quality in Radhanagar forest range under Bankura district. Eurasian J For Sci 8(1):1–10
Mandal M, Chatterjee ND (2020c) Elephant's habitat suitability assessment through geo spatial quantification in Panchet forest division, West Bengal. Ecofeminism Clim Change
Mandal M, Chattarjee ND (2020d) Geo-statistical analysis to understand nature of forest patch shape complexity in Panchet forest division under Bankura district, West Bengal. Indian J Ecol 47(1):96–101
Mandal M, Chettarjee ND, Hazra J (2015) Elephant migration and colonization in Bankura district, West Bengal, India. Vidyasagar University. Indian J Geogr Environ 14:46–52
McGarigal K, Marks BJ (1994) FRAGSTATS: spatial pattern analysis program for quantifying landscape structure. Version 2.0. Forest Science Department, Oregon State University, Corvallis, 67
Miller J, Rogan J (2007) Using GIS and remote sensing for ecological mapping and monitoring. Integr GIS Remote Sens 3:233
Mishra SR, Nayak AK (2010) Human-elephant conflict by inter-state migratory elephants (Elephas maximus) in Baripada & Balsore. Odisha, India
Moniruzzaman M, Bhowmick AR, Karan S, Mukherjee J (2021) Spatial heterogeneity within habitat indicates the community assemblage pattern and life strategies. Ecol Ind 123:107365
Naveh Z, Lieberman AS (2013) Landscape ecology: theory and application. Springer Science & Business Media
Pal M, Mather PM (2004) Assessment of the effectiveness of support vector machines for hyperspectral data. Futur Gener Comput Syst 20(7):1215–1225
Pal M, Mather PM (2005) Support vector machines for classification in remote sensing. Int J Remote Sens 26 (5):1007–1011
Pandit PK, Chanda S (2019) Human-elephant conflict and its possible control measures in South West Bengal Land Scape, India. Indian For 145(10):911–920

Pierik ME, Dell'Acqua M, Confalonieri R, Bocchi S, Gomarasca S (2016) Designing ecological corridors in a fragmented landscape: a fuzzy approach to circuit connectivity analysis. Ecol Ind 67:807–820

Puyravaud JP, Cushman SA, Davidar P, Madappa D (2017) Predicting landscape connectivity for the Asian elephant in its largest remaining subpopulation. Anim Conserv 20(3):225–234

Roy PS, Tomar S (2000) Biodiversity characterization at landscape level using geospatial modelling technique. Biol Cons 95(1):95–109

Roy A, Devi BSS, Debnath B, Murthy MSR (2010) Geospatial modelling for identification of potential ecological corridors in Orissa. J Indian Soc Remote Sens 38(3):387–399

Saaty TL (1999) Fundamentals of the analytic network process. In: Proceedings of the 5th international symposium on the analytic hierarchy process, pp 12–14

Santra AK, Samanta AK, Pan S (2007) Measures adopted to combat migratory elephants in South West Bengal forests. Gajah 27(2007):42–47

Satty T (1980) The analytical hierarchy process. McGraw Hill, New York

Shaw AK (2020) Causes and consequences of individual variation in animal movement. Mov Ecol 8(1):1–12

Shea K, Roxburgh SH, Rauschert ES (2004) Moving from pattern to process: coexistence mechanisms under intermediate disturbance regimes. Ecol Lett 7 (6):491–508

Singh AK (2006) Ecological Investigation of Human-Elephant Conflicts in South West Bengal, Doctoral dissertation, Saurashtra University

Singh AK, Singh RR, Chowdhury S (2002) Human-elephant conflicts in changed landscapes of south West Bengal, India. Indian For 128(10):1119–1132

Sukumar R (1990) Ecology of the Asian elephant in southern India. II. Feeding habits and crop raiding patterns. J Trop Ecol 33–53

Sukumar R (2003) The living elephants: evolutionary ecology, behaviour and conservation. Oxford University Press, New York

Taubert F, Fischer R, Groeneveld J, Lehmann S, Müller MS, Rödig E, Huth A (2018) Global patterns of tropical forest fragmentation. Nature 554(7693):519–522

Tong HL, Shi PJ (2020) Using ecosystem service supply and ecosystem sensitivity to identify landscape ecology security patterns in the Lanzhou-Xining urban agglomeration, China. J Mt Sci 17(11):2758–2773

Turner DP, Ollinger SV, Kimball JS (2004) Integrating remote sensing and ecosystem process models for landscape-to regional-scale analysis of the carbon cycle. Bioscience 54(6):573–584

van Toor ML, Kranstauber B, Newman SH, Prosser DJ, Takekawa JY, Technitis G, Safi K (2018) Integrating animal movement with habitat suitability for estimating dynamic migratory connectivity. Landscape Ecol 33(6):879–893

Yu H, Liu X, Kong B, Li R, Wang G (2019) Landscape ecology development supported by geospatial technologies: a review. Eco Inform 51:185–192

6 Wildlife Conflict Area Demarcation—Special Reference to Human–Elephant Conflict

Abstract

Human–elephant conflict (HEC) is presently a major issue in West Bengal. This chapter aims to demarcate conflict intensity zones based on selected landscape features, i.e. percentage of forest cover, multi-crop area density, built-up area density, forest habitat patch shape complexity road network buffering, canal buffering and water patch density. All of these data were obtained from individual forest range images (LISS-III, P6 Satellite image 2019). Individual thematic layers were prepared using kriging tool and reclassification done for five classes in ArcGIS 10.1 version software. The Analytic Hierarchy Process (AHP) method was used in order to elucidate weight and rank of selected individual parameters like forest cover, multi-crop area, built-up patch density, patch fragmentation, road and canal and water patch density. Weighted overlay analysis was applied to demarcate the wildlife conflict zones using selected parameter layers. This zonation will help the planners or policy makers for implementing preventive measures in controlling HEC effects.

6.1 Case Study from Bankura District, West Bengal, India

Habitat-based wildlife conflict zone mapping is in use for overall management, protection, conservation of target species (Ruda et al. 2018; Estes et al. 2012) and social awareness. Geospatial information helps to generate such kind of map when we considered large animal species in a specific landscape area. HEC is always landscape dependant ecological process (Woodroffe et al. 2005; Parker et al. 2007; Fernando et al. 2008). Landuse/land cover is an important factor for biodiversity and regional ecological function (Dickman 2010; Bogucki et al. 2012). In a forest landscape, animal movement processes are controlled by adjoining other landuse pattern. Therefore, landuse change influences the ecological behaviour of animal (Desai and Hedges 2010; Chanda 1996).

Similar tendency is found in the study area of Bankura district. Quality of forest quality on this district improved with reference to amount and health (Kulandaivel 2010; Chatterjee 2016). The substantial landuse changes are found here since 1980 after successful implementation of Joint

M. Mandal and N. Das Chatterjee, *Geo-Spatial Analysis of Forest Landscape for Wildlife Management*, GIScience and Geo-environmental Modelling, https://doi.org/10.1007/978-3-031-33606-5_6

Forest Management (JFM) and several forest development programs. This changing nature of landscape immensely changed the regional biodiversity process. HEC process is one of them. It becomes regular and causes human death, injury, crop damage, property loss and even death of animal. Human death by elephant is a common incident in the district. In fact, the study area is the part of extended home range of Dalma's elephant, Jharkhand State (Sukumar 2003; Singh et al. 2002; Singh 2006). Since two decades, they become colonized in the forests of Bankura supported by regenerated forest patch (Chatterjee 2016; Mandal et al. 2015) and proximate agricultural land an accessible food source (Mandal and Chatterjee 2018). Before 1987, elephant related activities were not common but occasionally found (Chowdhury et al. 1997; Chanda 1996). It is interesting to know that HEC frequency gradually increases and now becomes a serious issue in social process (Chatterjee 2016; Mandal and Chatterjee 2020b). It is found that the maximum HEC event concentrated in specific villages under the study area. Geospatial factors are responsible for spatial concentration of HEC. Location of the forest habitat may be the leading cause of this fact, but other factors like anthropogenic intrusion, habitat fragmentation, water source, human settlement and crop area inside the habitat are also responsible. Elephants shared these landscapes for their biological interaction. Hence habitat preference or selection are found with these landscapes.

Disturbance is another factor for increasing HEC (Desai 2008; Sukumar 1994). Elephants usually avoid disturbed areas. Thus there is high chance of HEC in the highly disturbed areas (Forman 2012). Road, built-up area, fence, canal are the common disturbed landscape elements which divert the movement pattern of wildlife (Kumar et al. 2010). As result of it, huge conflict is found in these areas. And the risks of conflicts are found minimum when the distance increases from these disturbed landscapes (Cushman et al. 2010). Therefore, it seems that landscape features control nature of HEC in a region (McGarigal and Marks 1995). Considering all these landscape factors, a conflict risk zone can be predicted using spatial association and interpolation which demarcates the location of conflicts.

The probable HEC risk zone has a multidimensional impact. Different issues like risk minimization, awareness generation, management may be taken on the basis of this zonation. But HEC zonation cannot be understood without spatial mapping and geostatistical analysis (Miller 2015). For this, the present study tries to delineate HEC risk zone by selected landscape features (parameter) in Bankura district through weighted method using Analytic Hierarchy Process and weighted overlay analysis. The final output will help to implement such kind of approaches which controls the HEC in the study area.

6.1.1 Methods and Data Used

Bankura district is region with enormous geographic significance, located at the intermediate zone between Chhotonagpur plateau and the plains of Bengal. It lies between 20° 38′ 01.09″ N to 23° 38′ 06.24″ N and 86° 36′ 18.35″ E to 87° 46′ 25.56″ E. As per West Bengal State Forest Report 2017 the geographical area of the district is 6882 km^2 with 21.53% forest cover which is the highest among all districts of western part of West Bengal. Agriculture holds the dominant landuse/land cover, i.e. 59% out of total area. Paddy and vegetables are the major crop produced using the canal irrigation in the east. But in the western portion of the district only Kharif crop is practiced. The second dominant landuse is forest cover. As result, wilderness and wildlife conflicts are common in this district specially HEC (Mandal and Chatterjee 2020a). Bankura district forest administration is hierarchically arranged in Division, Range and Beat. Three Forest divisions are hierarchically ordered into 28 Ranges and 93 Beats (Fig. 6.1).

The landscape attributes are considered as parameters for assessing HEC. IRS P6 LISS-III satellite image 2017 of Bankura district is collected from ISRO Departmental website. This

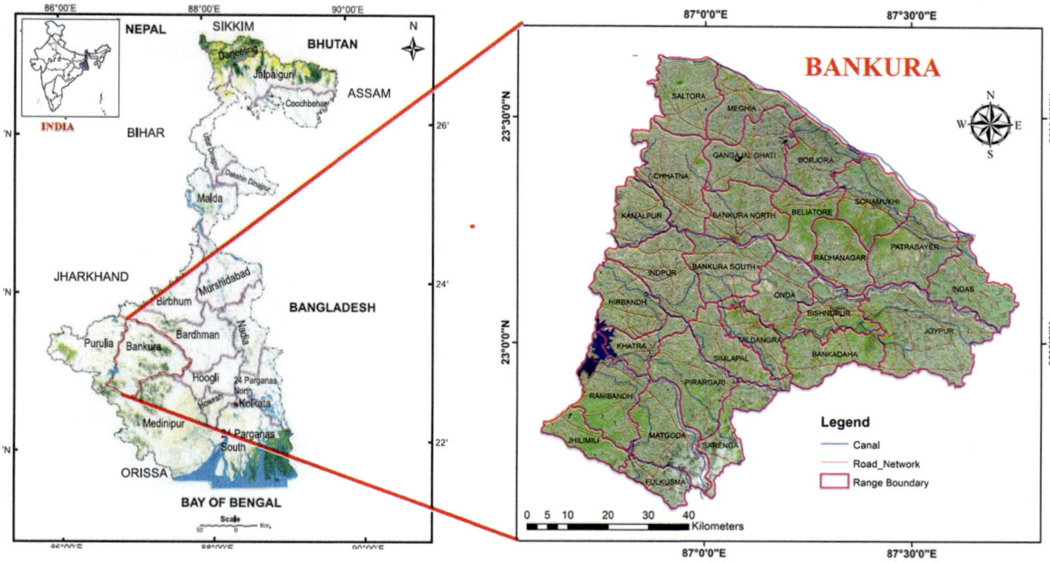

Fig. 6.1 Location map of the study area with major roads, canals and forest range boundary

image is classified by unsupervised method through ERDAS 9.3 and ArcGIS 10.3 version software. Forest cover, built-up area, water point and multi-crop area maps are separated individually from this classified LULC image. Forest range-wise individual parameter values (amount and percentage) were calculated. Then forest range-wise AWMSI was measured to know the nature of forest fragmentation for wildlife conflict zone demarcation (Dramstad et al. 1996). Finally, these five categorical forest range-wise values converted into point feature to get density distribution maps in raster format. The other two parameters, i.e. road and canal were considered and digitized from Google image 2018. After that road and canal buffer maps were prepared.

6.1.2 Measuring Landscape Unit

Landscape is the container for any kind of ecological process (Forman 1995). Habitat function is also controlled by landscape attributes. Wildlife conflict is one of the most important ecological events which depend on landscape function and utilization. The landuse/land cover plays a great role in ecological process like species migration, movement, colonization, foraging (Dramstad et al. 1996) etc. especially for elephant (*Elephas maximus*). All these processes are common and responsible for HEC. At the time of movement from one habitat patch to nearest similar habitat they face difficulties due to presence of cultural landscape like built-up areas, canal, road, etc. Another landuse factor is crop type and pattern. Herbivore animals like elephant always tries to colonize where they get food easily throughout the year (Sukumar 2003). Therefore, HEC is common where species colonizing landscape factors and disturbing factors are densely present. Based on this, the present study considers five landuse pattern density (Figs. 6.2, 6.3, 6.4, 6.5, and 6.6) and two landuse buffering (Figs. 6.7 and 6.8) which influence wildlife conflict (HEC) areas.

6.1.2.1 Forest Cover

Wildlife habitat is directly linked with forest cover. Nature of the forest habitat supports species ecological several functions like rest, hide, food, disturbances etc. Forest in the study area becomes isolated and patchy in nature (Chatterjee 2016; Singh 2006). Therefore, elephant forced to stay where forest become dense and

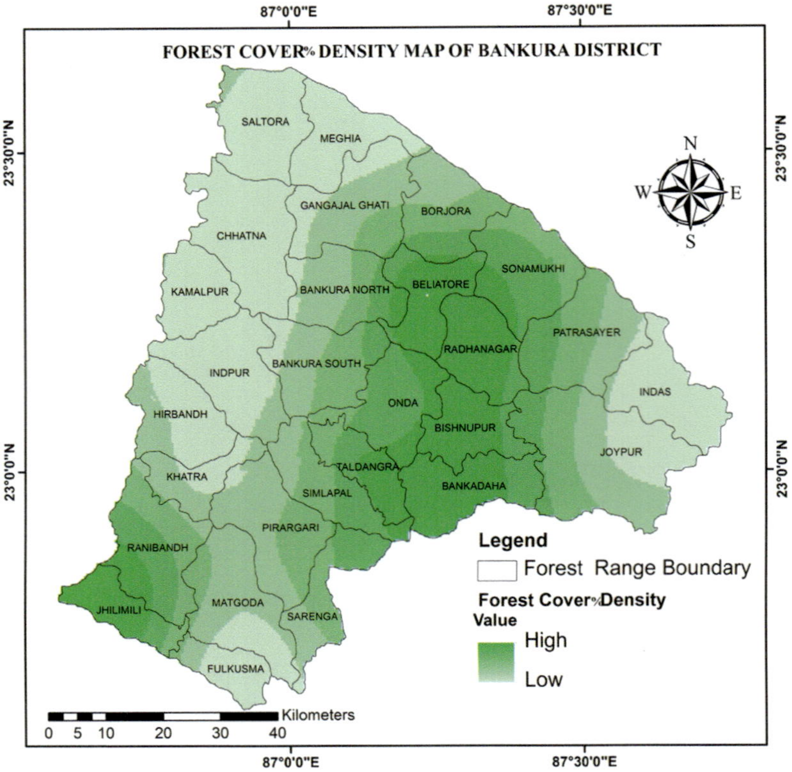

Fig. 6.2 Forest range-wise forest cover density map of Bankura district

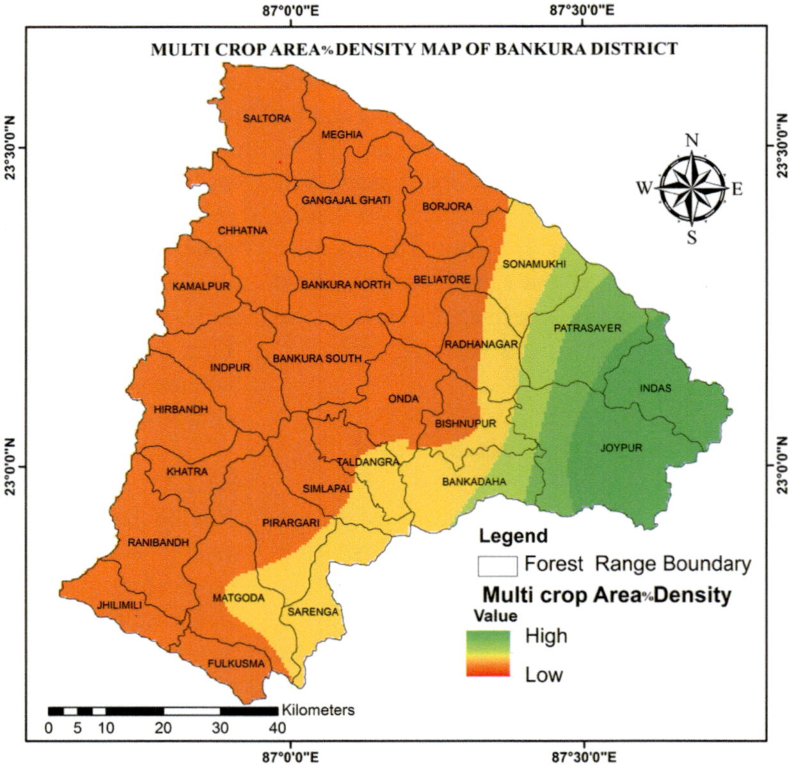

Fig. 6.3 Forest range-wise multi-crop area density map of Bankura district

Fig. 6.4 Forest range-wise AWMSI density map of Bankura district

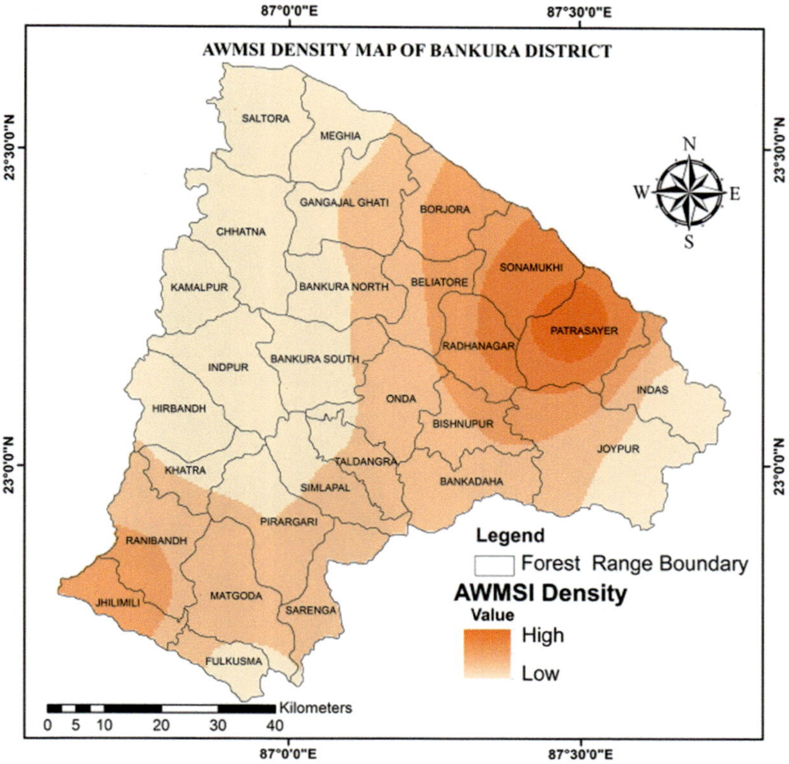

Fig. 6.5 Forest range-wise built-up patch density map of Bankura district

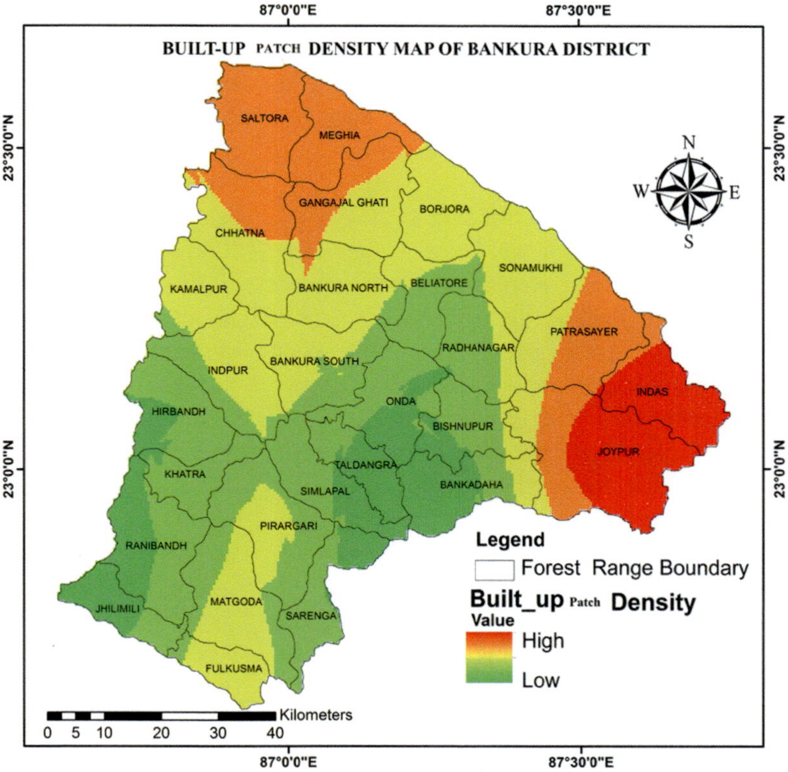

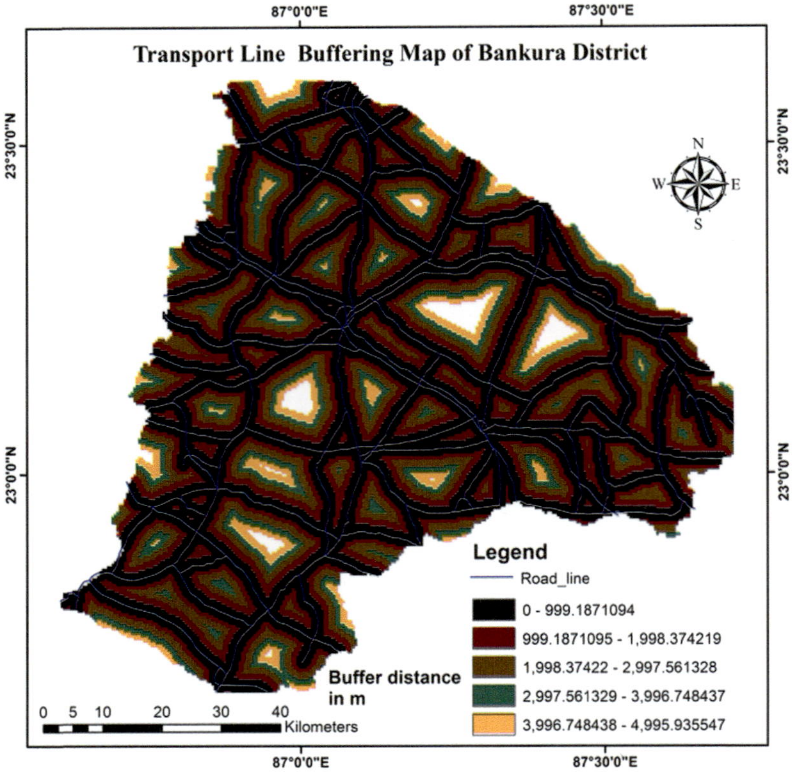

Fig. 6.6 Classified map of buffering transport line under Bankura district

highly proximate. For that reason, the study considered forest cover density as a factor to detect conflict area because in these areas elephant movement become frequent than less dense forest cover areas.

6.1.2.2 Multi-crop Area

Food accessibility is a big factor for any large herbivores. In Africa, elephants have to migrate miles to take food at the time of dry season (Viljoen 1989; Anthony and Avery 2008). Food selection and amount of food availability are two major factors of animal ecology (Pokharel et al. 2018). Elephants like palatable secondary vegetation for their forage than natural vegetation. They used to stay in the forests near the agricultural fields for easy accessibility of crops and vegetables (Santra et al. 2008). Therefore, multi-crops near the forest fringe areas are the suitable location for elephant staying (Panja and Mistri 2018; Pradhan and Wegge 2007). These areas will be under high HEC due to frequent movement from forest to feeding ground.

6.1.2.3 Built-Up Area

The present study considers settled and industrial area as built-up area. The scatter settled areas near the forest habitat are more prone to conflict (Forman 2012; Dramstad et al. 1996; Von Gerhardt et al. 2014). Bankura district is the part of extended home range of Dalma elephant (Chowdhury et al. 1997; Datye and Bhagwat 1995; Sukumar 2003) thus they roam at the adjoining forest patches frequently. At the time of their movement, they faced human settlements situated in between forest patches or across their corridors. So where patch density of built up area is high, there conflict intensity is also high (Chakraborty and Mondal 2013).

6.1.2.4 Forest Fragmentation—Area Weighted Mean Shape Index (AWMSI)

Forest fragmentation promotes edge contrast (Farina 2008; Forman 1995) or forest shape complexity. HEC will be high when edge contrast is high. Forest fragmentation due to

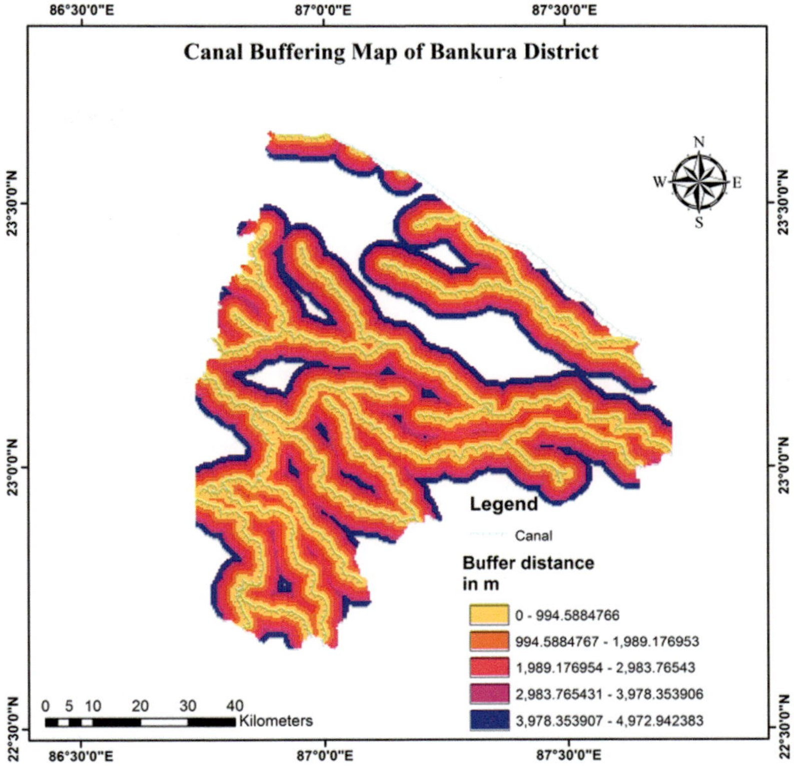

Fig. 6.7 Classified maps of canals (buffering at different distance) under Bankura district

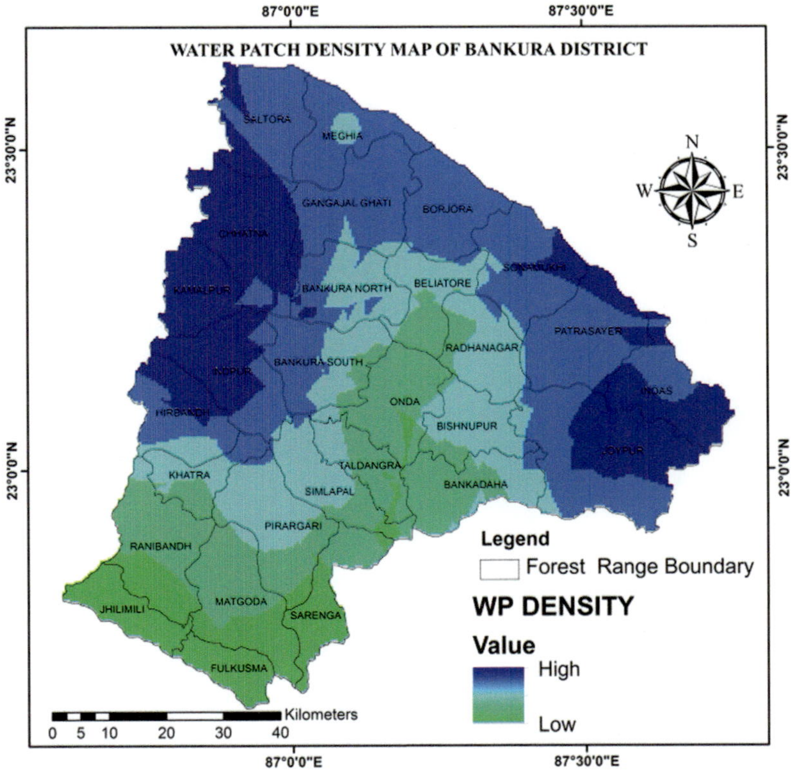

Fig. 6.8 Forest range-wise water patch density map of Bankura district

intrusion of anthropogenic activities like agricultural expansion, industrial development, settlement built-up, network line development etc. creates habitat shape complexity (Li and Wu 2007). Highly complex forest shape become more prone to conflict (Couvillion 2005). Where forest becomes more fragmented, conflict will be high (McGarigal and Cushman 2005). To measure this forest habitat shape complexity AWMSI is applied using following (McGarigal and Cushman 2005) method. The index value when high complexity is also high. This index is calculated for the forest ranges using FragStat 4.2 version software and density map is prepared.

$$\mathrm{AWMSI} = \sum_{j=1}^{n}\left[\left(\frac{0.25p_{ij}}{\sqrt{a_{ij}}}\right)\left(\frac{a_{ij}}{\sum_{j=1}^{n} a_{ij}}\right)\right]$$

p_{ij} = perimeter (m) of patch ij
a_{ij} = area (m^2) of patch ij
n_i = number of patches in the landscape of patch type (class) i.

6.1.2.5 Road Network

Anthropogenic intervention on natural landscape always disturbed the free movement of animals. Road or any kind of transport lines make a barrier for large size animal's movement in their home range (Forman and Deblinger 2000). In Bankura district, transport lines are densely distributed and crossed many forest core areas. Sometimes elephants died by road accident and human death are also common near the forest road. So the distance from the road is a factor to understand the conflict. There is a probability that long the distance from road (Forman 2012). Therefore, road line buffering thematic map is prepared to demarcate the probable conflict zone in Bankura district.

6.1.2.6 River and Canal

River or canal is a big obstacle for the movement of large herbivores (Forman 1995). So the river side is highly prone to HEC because it creates barrier. It also hampers regular movement of migrated elephant. Similar method is followed similar to road buffering to determine the probable conflict zone in case of river and canal.

6.1.2.7 Water Patch

Water is very essential for species survival. Elephant's water requirement is huge in nature. Elephants migrate several miles within a day only for drink water in dry season (Loarie et al. 2009). So the source of water point must be a conflict area for its basic requirement. If water point density becomes less, then conflict will be high because lack of water points diverts the movement of elephants which led more HEC. Therefore, to understand the conflict zone Water Patch Density (WPD) map is prepared.

6.1.3 Taking Analytic Hierarchy Process (AHP) Approach

The Analytic Hierarchy Process is a multi-criteria decision-making method established by Saaty (1977, 1994). In the AHP method, a hierarchical model consisting of objectives and criteria is used for every problem (Saaty 1990, 2004). After establishing the problem on a hierarchical structure, the weights of the criteria are forming the hierarchy (Öztürk and Batuk 2010). To evaluate the criteria compared to other criteria included in the next hierarchy level. An individual criterion is scored with the use of the favourite modified scale (Table 6.1) recommended by Saaty (1980). The pair-wise comparison matrix consists of $n (n - 1)/2$ comparisons for 'n' number of elements (Malczewski 1999; Mokarram and Aminzadeh 2010; Akinci et al. 2013). Weights or priorities are determined by normalizing the matrix. For this normalization, a "normalized pair-wise comparison matrix" is calculated dividing the column elements of the matrix by the sum of each column. The row elements in the calculated matrix are summed, and the total value is divided by the number of elements in the row. In this manner, a weight vector is gained (Mendas and Delali 2012). Weights are within the range of 0–1, and their sum is equal to 1

Table 6.1 Scale of weighting parameters in comparison matrix after modification of Saaty (1980)

Level of importance	Value mean for
1	Equal importance
3	Weak importance of one over another
5	Essential or strong importance
7	Demonstrated importance
9	Absolute importance

(Malczewski 1999). Pair-wise comparison matrix in AHP consistency is very important (Öztürk and Batuk 2010). To measure this consistency, consistency ratio (CR) by Saaty (1980) technique is used. The matrix is acceptable when CR value is less than 0.10.

A comparison matrix is done for getting the weight and rank of individual selected parameters following AHP method (Table 6.2). The compression judgement is carried out by several expert opinion and field investigation. The present comparison matrix is continued because CR is 0.0726 which is less than '0.01' that assessed significant degree of acceptance.

These prepared thematic raster layers then classified individually with equal interval in considerable subcriteria. Subcriteria of parameters are also scored within the rank 1–5 against each class (Table 6.3). In this scoring of subcriteria, high score, i.e. 5 is highly influenced HEC and low, i.e. 1 is insignificant for HEC.

After parameter weights and subparameter individual scores, prepared raster map of seven parameters are overlaid using the weighted sum

Table 6.2 Pair-wise comparison matrix for parameters weight

Matrix		Forest cover	Multi-crop area	Built-up area	AWMSI	Road	Canal	WPD	Normalized priority-vector	Weight in %	Rank
		1	2	3	4	5	6	7			
Forest Cover	1	1	1	3	3	5	7	9	0.327818	32.78%	1
Multi-crop area	2	1	1	1	1/3	3	5	7	0.177432	17.74%	3
Built-up area	3	1/3	1	1	1/3	3	3	3	0.12139	12.14%	4
AWMSI	4	1/3	3	3	1	3	3	5	0.227615	22.76%	2
Road	5	1/5	1/3	1/3	1/3	1	3	3	0.070737	7.07%	5
Canal	6	1/7	1/5	1/3	1/3	1/3	1	3	0.046644	4.66%	6
WPD	7	1/9	1/7	1/3	1/5	1/3	1/3	1	0.028364	2.84%	7
									$\sum = 1$	$\sum = 100$	

Max. eigenvalue (y_{max}) = 7.588
$n = 7$
Consistency Index (CI) = $(y_{max} - n)/(n - 1) = 0.098$
Random Index (RI) = 1.35
Consistency Ratio (CR) = CI/RI = 0.0726

Table 6.3 Weightage for parameters subcriteria for weighted overlay analysis to detect wildlife (HEC) conflict zone

Selected parameters	Weight in %	Parameter subcriteria classes	Score
Forest cover % density	32.78	< 7.236	1
		7.236–14.375	2
		14.375–21.514 in %	3
		21.514–28.652	4
		> 28.652	5
Multi-crop area % density	17.74	0.086–10.136	1
		10.136–20.185	2
		20.185–30.234 in %	3
		30.234–40.283	4
		40.283–50.332	5
Built-up patch density	12.14	5.608–6.070	1
		6.070–6.532	2
		6.532–6.994/ha	3
		6.994–7.456	4
		7.456–7.918	5
AWMSI density	22.76	2.974–4.889	1
		4.889–6.804	2
		6.804–8.719–	3
		8.719–10.634	4
		10.634–12.549	5
Road buffering	7.07	< 1	5
		1–2	4
		2–3 km	3
		3–4	2
		> 4	1
Canal buffering	4.66	< 1	5
		1–2	4
		2–3 km	3
		3–4	2
		> 4	1
WPD	2.84	0.724–0.783	5
		0.783–0.843	4
		0.843–0.902/ha	3
		0.902–0.962	2
		0.962–1.021	1

overlay analysis tool in ArcGIS 10.1 version software, and finally, wildlife conflict risk zone map is created. The output layer is also divided into four classes of equal ranges on the basis of conflict risk intensity. The total working method is continued sequentially to get the results (Fig. 6.9).

6.1.4 Results

After technical analysis of geospatial factors, the final map comes out (Fig. 6.10). In this categorical map, four conflict risk zones are demarcated as less conflict zone, medium conflict zone, high conflict zone and very high conflict zone. The very high conflict zone is 1064.50338 km^2 covered the forest ranges like Bankadaha, Borjora, Patrasayer Bishnupur, Radhanagar, Joypur, Beliatore and Sonamukhi (Table 6.4).

The middle portion of the district is medium to high risk zone for wildlife conflict. The area under high conflict zone is 2101.46 km^2. It covers the forest ranges like Indas, Jhilimili, Ranibandh, Sarenga, Taldangra, Borjora and Onda. Medium conflict zone is extended for 2555.40 km^2 area covered the forest ranges like Gangajalghati, Bankura North and South,

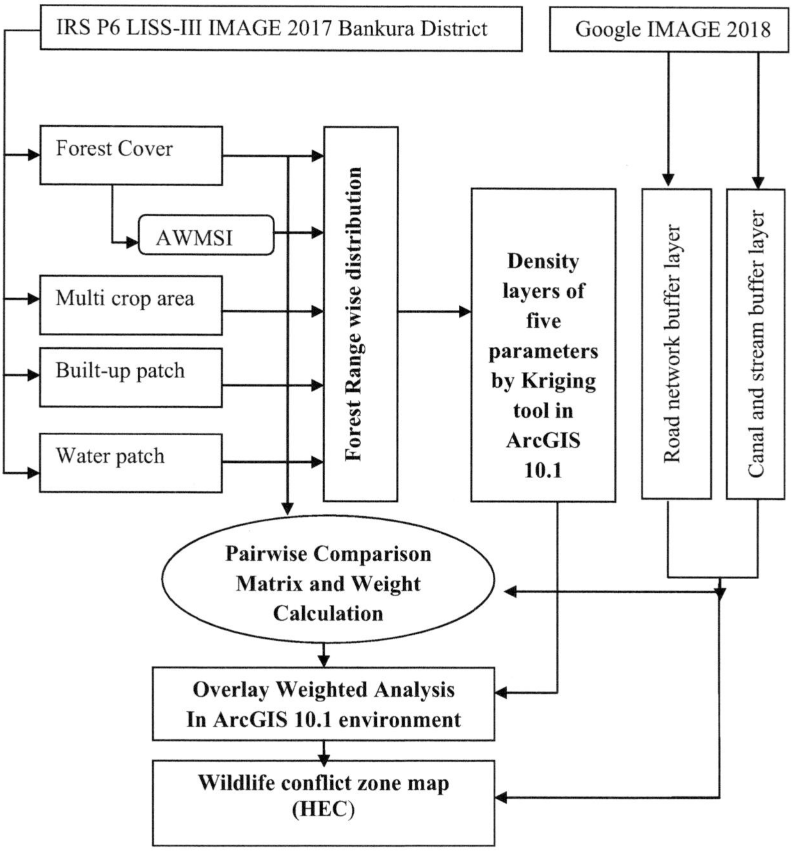

Fig. 6.9 Research methodological outlook for understanding the whole work

Simlapal and Matgoda. Area under less conflict zone is 1012.20 km^2 covered the forest ranges like Saltora, Chhatna, Hirbandh, Indpur, Kamalpur, Mejhia and Fulkusma where forest cover is minimum and less amount of multi-cropped area have become unfavoured for elephant frequent movement.

After getting the final map, it is found that the two major roads (SH-2 and NH-60), railway lines (Kharagapur to Adra and Bishnupur to Arambag) and Kansabati canal cross through the middle portion of the very high conflict zone (Fig. 6.11). Another observation is that very high conflict zone is enclosed by those forest ranges which have high density of forest cover and multi-cropped areas. In such very high conflict areas, human settlement units (built-up areas) are scattering in nature except Joypur and Bishnupur forest range.

6.1.5 Conclusion

The dynamic characters of HEC affect both human society and animal communities. Death of human being by wild animal is an accident likewise wild animal death by human is a big loss too. HEC issues like crop damage, human death and injury, hut damage, death of elephant, etc., are found more frequently in a very risk zone (Mumby and Plotnik 2018). According to forest stuffs of Bankura North and Panchet forest division under Bankura district, the most affected villages are Marar, Saltora, Punchmura, Kalabagan, Peardoba, Belsulia, Kantaberia, Nakaijuri, Kuchiakol, Salda, Hetla, Bhora, Kushdwip, Birsingha, Panchal, Kustal, Chandra, etc. (Figs. 6.12 and 6.13). They also informed that these villages are prone to HEC, so they give more attention to control HEC by frequent

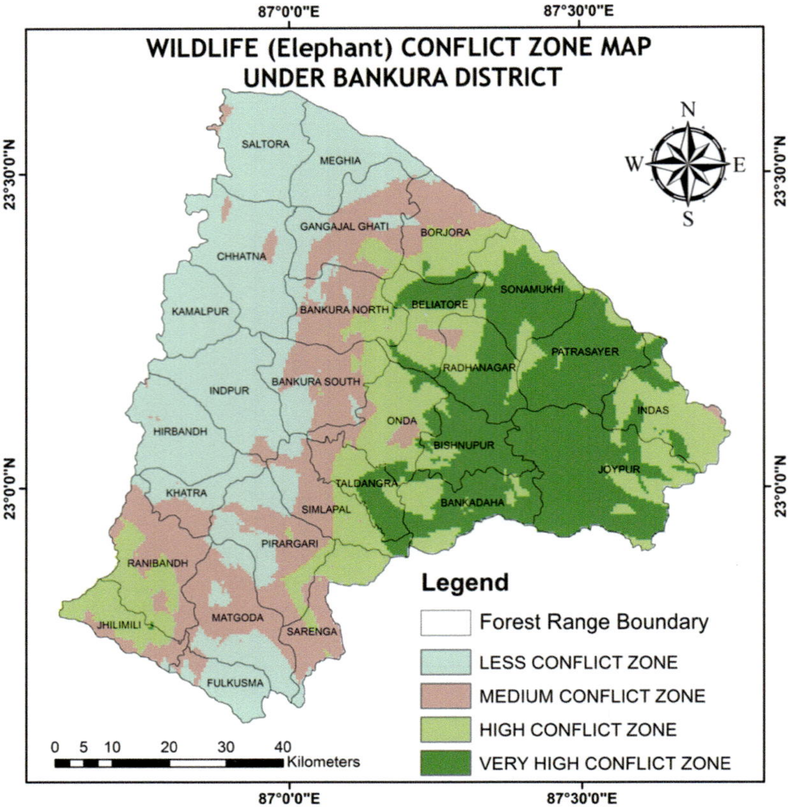

Fig. 6.10 Forest range-wise wildlife conflict vulnerability map of Bankura district

Table 6.4 Conflict zone wise area and forest ranges under Bankura district

Zone	Area in km^2	Forest ranges
Less conflict	1012.20869097	Saltora, Mejhia, Chhatna, Kamalpur, Indpur, Hirbandh
Medium conflict	2555.404048	Gangajalghati, Bankura North and South, Simlapal, Matgoda,
High conflict	2101.46789522	Indas, Jhilimili, Ranibandh, Sarenga, Taldangra, Onda, Borjora
Very high conflict	1064.50338188	Beliatore, Sonamukhi, Radhanagar, Patrasayer, Bishnupur, Joypur, Bankadaha

patrolling and forecast exact location of elephant herd by web media. The present study proves that these villages are facing maximum HEC because the, habitat conditions are better and disturbances are high. In medium or low conflict zone, only Kharif cultivations are practiced during monsoon season. As a result, elephant herd move occasionally. Human death or injury and even elephant deaths may be recorded in these areas.

It is interesting to know that elephants become residential in very high risk zone depending on suitable habitat condition (Chatterjee and Chatterjee 2014; Mandal and Chatterjee 2018). This fact further promotes HEC. Elephant death caused by electric fencing and train collision may be found in the particular highly conflict zone. In 2016, three elephants were killed and two injured by train collision in between Peardoba and Bishnupur station, and two elephants were killed

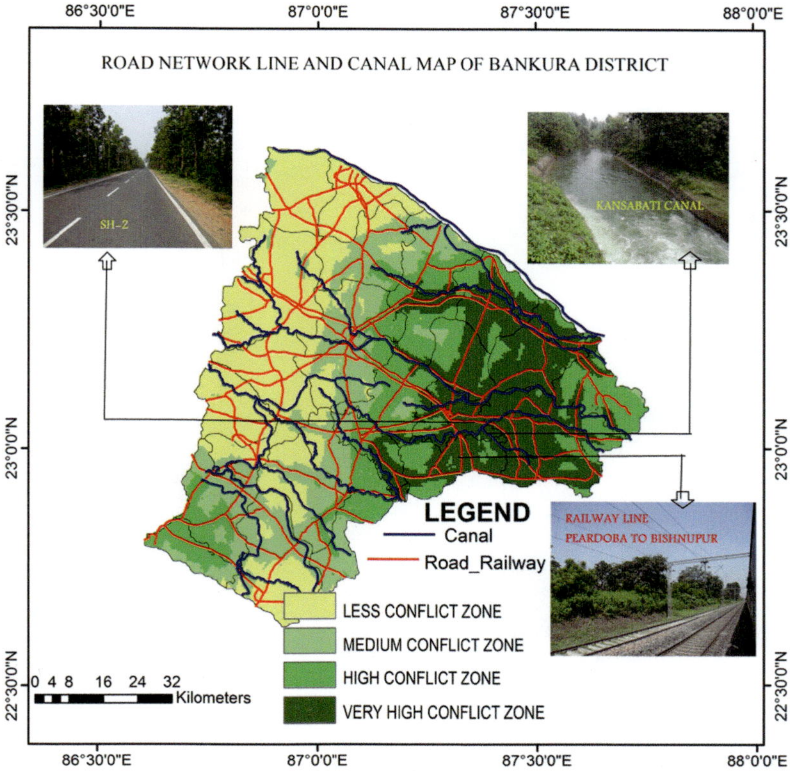

Fig. 6.11 Road and canal passes through conflict risk zone in Bankura district

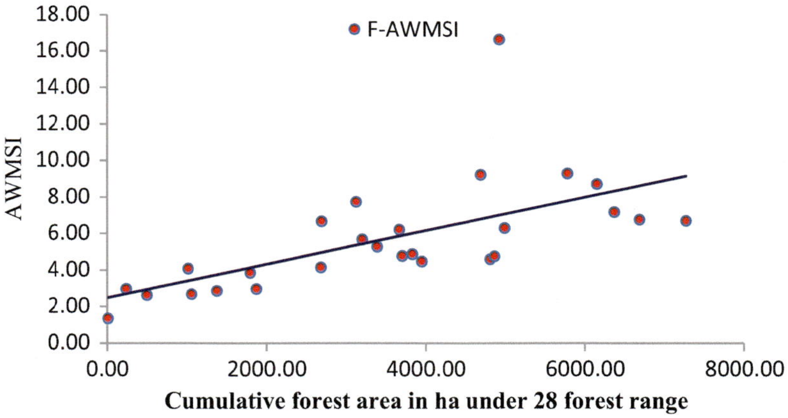

Fig. 6.12 Relation between forest area and forest fragmentation by AWMSI under 28 forest ranges in Bankura district

by electric fence in Sonamukhi and Beliatore forest range (Forest report 2017). Both these two cases, the location is under in very high conflict zone.

These incidents or accidents are found or happened where forest cover is fragmented in this district. Ecologically, amount of forest cover always control species habitat interaction (McIlroy 1978; Holbrook et al. 2019). Elephant herd in this region always attempt to stay where forest cover is better with proximate agricultural land and water (Chatterjee and Chatterjee 2014; Mandal and Chatterjee 2018) and forest fragmentation. Fragmented forest forced wildlife for

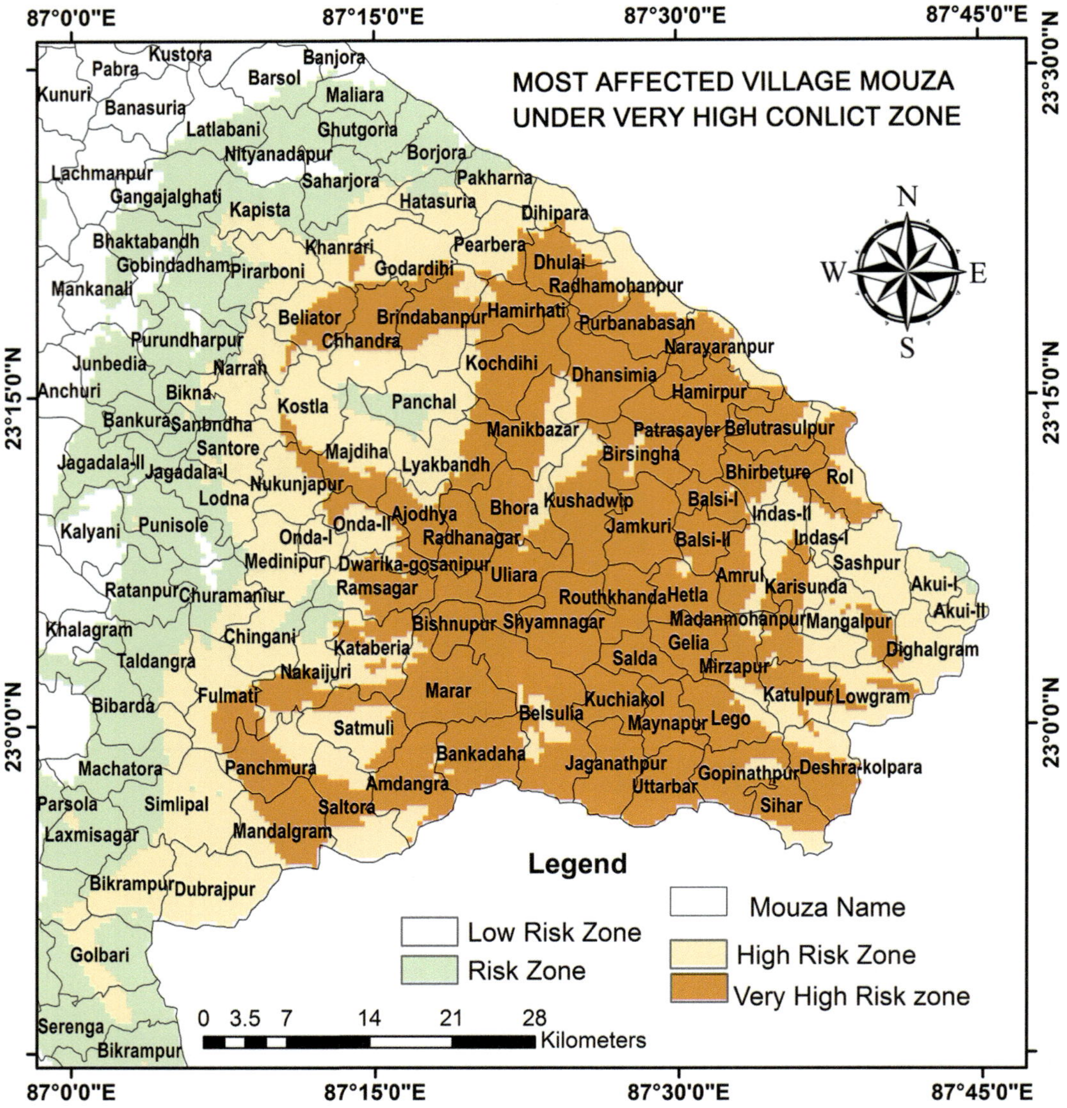

Fig. 6.13 Most affected village mouzas by HEC under Bankura district

frequent movement (Li and Wu 2007; Fahrig et al. 2019). It is found that for a short period of time, forest fragmentation increases with increase of forest area under 28 forest ranges in Bankura district (Fig. 6.12). This situation is one of the causes of high risk for wildlife conflict in these forest ranges like Sonamukhi, Beliatore, Bishnupur, Taldangra, etc. Other factor is multi-cropped area. Multi-crop offers accessible and available food throughout the year for elephant in this forest ranges. Multi-crop cultivated areas in these forest ranges are exposed to frequent and huge amount of damage by wild elephant. The main cause is that elephant preferred this secondary vegetation for their day to day diet than natural vegetation (Hill 2018). Fewer number of water bodies are also important factor for conflict between human and elephant (Mumby and Plotnik 2018) in these forest ranges. Scatter settlements suffer maximum conflict by elephants

(Gaynor et al. 2018) in the very high risk zone though built-up areas (settlement) density is moderate. NH-60 and State Highway-2 passes through the middle of the Joypur and Peardoba forest, i.e. a part of very high conflict zone where elephant's road crossing event are regularly found.

Ultimately, the method of the present study linked with geospatial attributes and elephant behaviour. Without any spatial information any kind of wildlife management program could not be conducted. So the findings of the present study have a significance to manage HEC. The dynamic nature of HEC concentration in Bankura district is analysed elaborately. Anthropogenic intrusion into the wild habitat is a significant cause of HEC that has been evident from this analysis. It is again a fact that biogeographical process directly linked with geospatial attributes in a landscape. Therefore, landscape mapping is very appropriate approach to measure the function of biogeographic process (Kupfer 2012). Another significance of the present study is that the study represents the predictive map (conflict risk map) on the basis of geospatial information. So the scientific management of these habitats in the present landscape can control the overall HEC situation in this district.

6.2 Case Study from Dalma Wildlife Sanctuary, Jharkhand, India

In the beginning of twenty-first century, GIS statistics and software modelling analysis play a significant role in wildlife habitat management (Farina 2008; Reza et al. 2013; Imam and Tesfamichael 2013). GIS is a powerful set of tools used to collect, store, retrieve, transform and present spatially referenced environmental data from the real world (Burrough and Mcdonnell 1998). It is now used for a wide range of applications for answering questions on the ecology (Akcakaya 2000) and distribution of individual species and communities (Vogiatzakis 2003). Now it is received a lot of attention in the field of ecology (Forman 1995; Kushwaha and Roy 2002) in recent years. Here, it has been used in landscape ecology and ecosystem research and has gradually extended to the field of individual and behavioural ecology (Liu et al. 1997). The present study has also committed with this application to measure habitat suitability for addressing HEC nature in DWS.

The habitat analysis is considered as most important in planning and management of protected area (Cushman 2006; Wiersma et al. 2011). Asian elephant (*Elephas maximus*) is recognized as an endangered species by Convention on International Trade in Endangered Species of Wild Fauna and Flora (Hedges and Gunaryadi 2010). Recently, elephants in DWS is facing a major problem that led HEC (Chatterjee and Chatterjee 2014; Ranjan et al. 2016). Increased human interference in landuse alteration resulted isolation of elephant populations in fragmented areas (Sukumar 2003; Desai and Hedges 2010). Habitat loss and fragmentation due to human interference has become a huge threat to their survival process (Kanagaraj et al. 2019). As a result of it, elephants dispersed from core habitat to the surroundings (Baskaran et al. 2011). The alteration of physical attributes in DWS is responsible (Das et al. 2018). So HEC characters are multidimensional in nature that would be understood by an inter-disciplinary approach where the concepts of geography, ecology and wildlife biology are acting together.

From the literature it may be said that habitat of elephant completely depend on geospatial attributes. Elephants wander in search of food, water and possible mates from one place to other place (Guha 2017; Sitompul et al. 2013; Areendran et al. 2011). To do so, they generally follow a path. Landscape character determines their choice of resting, foraging and shelter between the places. Therefore, geospatial modelling is an useful techniques to identify the potential elephant habitat (Bist 2006) in DWS. The landscape attributes like rivers, water holes, favourable vegetation type which offer forage, etc., act in favour of elephant habitat while roads, settlements, steep slope, etc., offer a resistance to their free movement. All of these mentioned spatial attributes are taken as input variable to create an elephant habitat suitability map by AHP and overly analysis method.

The present study tries to address the HEC situation on the basis of habitat suitability zone. Generally, highly suitable elephant habitat areas will be the intense affected (Desai and Riddle 2015) and less suitable areas will be fewer affected of HEC. The causes of HEC are different in different situation and location. Sometimes dominant suitable habitat factors may not be act as a controlling factor for HEC. Landuse practices have a strong influence on HEC character (Chatterjee 2016; Mandal et al. 2015). This ecological trend has a background or land mosaic effect that would be understood after habitat suitability zoning in DWS. Multi-habitat function merely controls wild ranging animal movement in their home range. Such elephant movement pattern will lead HEC in different places under DWS.

6.2.1 Materials and Methods

Jharkhand or the land of forest is a state in eastern India. This state carved out from the southern part of Bihar on 15 November 2000. DWS is a unique biogeographical comes under biotic province of Chhotanagpur (6B) of Deccan peninsula zone within the Indomalayan region in Jharkhand (Rodgers and Panwar 1988). DWS lies between 22° 45′ N to 23° N and 86° 0′ E to 86° 25′ E covering a surface area of 464.01 km^2 (Fig. 6.14).

The sanctuary is comprising an area of 193.22 km^2. It is the part of a hill range running South East to North West and traditionally called the Dalma hill range. The whole range has several peaks having different local names. The highest point of the range in Dalma hill is 926 m above M.S.L. and local name is Sadhubera Pahar. The sanctuary encloses 29 revenue villages within its limit, and another 73 villages exist along the periphery. The township of Jamshedpur is only two kilometre away from the nearest point of the boundary of the sanctuary.

6.2.1.1 Data Base and Data Source

To execute the present study and to fulfil the objectives different data sets are used both primary as well as secondary raster and vector data. Survey of India topographic maps with a scale of 1:50,000 and series No. F45I5.73j/5, F45I1 and 73j/1 are used to verify the landsue land cover classification. Essential parameters are used in the study-elevation, slope, hill shade and aspect was derived from the CARTOSAT DEM (30 m) that was obtained from BHUVAN, National Remote Sensing Centre, India. ERDAS 9.3 imagine, ArcGIS 10.3 and ENVI software was used for digital image processing and statistical spatial analysis. Landsat-8, OLI 2018 high resolution multi-spectral data were interpreted using supervised classification techniques to perform the landuse land cover map preparing. Landsat image was obtained from Earth Explorer, United States Geological Survey (USGS) site (http://glovis.usgs.gov/). Satellite images have been classified into seven landuse land cover classes taking MINIMUM DISTANCE algorithm using ERDAS IMAGING 9.3 software. These seven classes are dense forest, open forest, water body, settlement, agriculture land, fallow land and wet land (Fig. 6.15). In 2018, total landuse and land cover area is 473.59 km^2 where dense forest, agricultural land, open forest, settlement, water body wet land and fellow land covered 33%, 31%, 15%,11%, 5%, 3% and 2%, respectively.

6.2.2 Selected Criterion Maps for Elephant Habitat Suitability

Selected landuse factors are used for preparing the suitable habitat areas for elephants. These criterion maps are prepared for the demand of weighted overlay analysis in GIS environment following cumulative steps (Fig. 6.16). It helps to make the decision of suitable habitat areas of elephant in DWS. The criterion map is classified

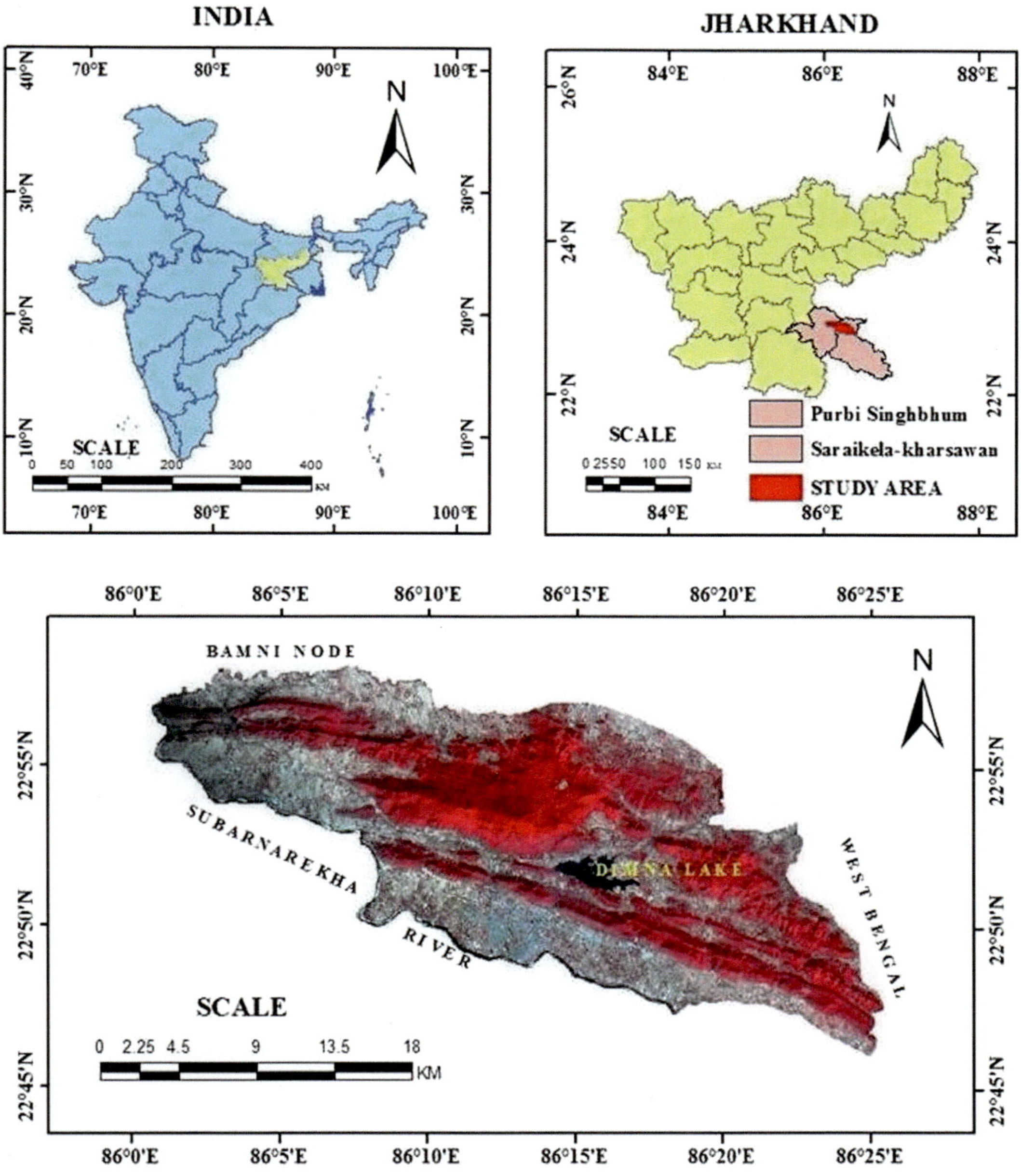

Fig. 6.14 Location map of Dalma Wildlife Sanctuary in Jharkhand State, India

in equal interval classes. Single class represents the density range that has a significance positive or negative impact in elephant habitat suitability function.

6.2.2.1 Forest

Forest composition is one of the important criteria of elephant's habitat suitability analysis. It is most important factor for elephant habitat because one adult elephant eat 150–170 kg vegetation daily (Sukumar 2003) form of leaves, twigs, fruit, bark, grass and roots. Forest also gives shelter elephant. Densely forest has a good capacity to enrich habitat quality. Forest class map was prepared by unsupervised method with distinct three classes that represents the total area

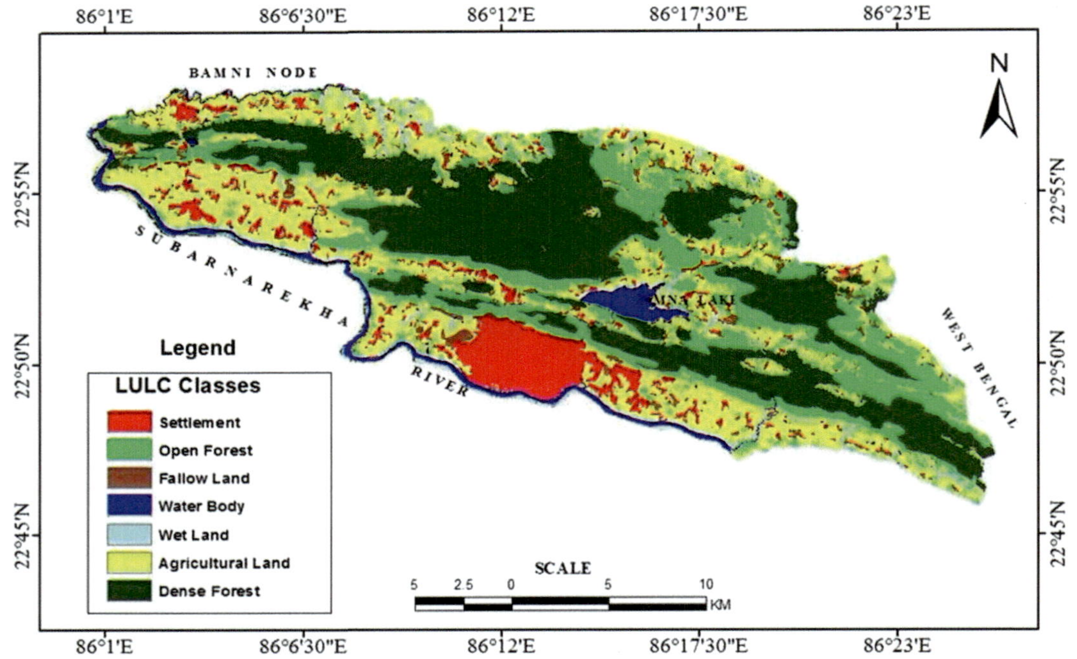

Fig. 6.15 Map of Landuse land cover classes in DWS, 2018

Fig. 6.16 Research deign of the work

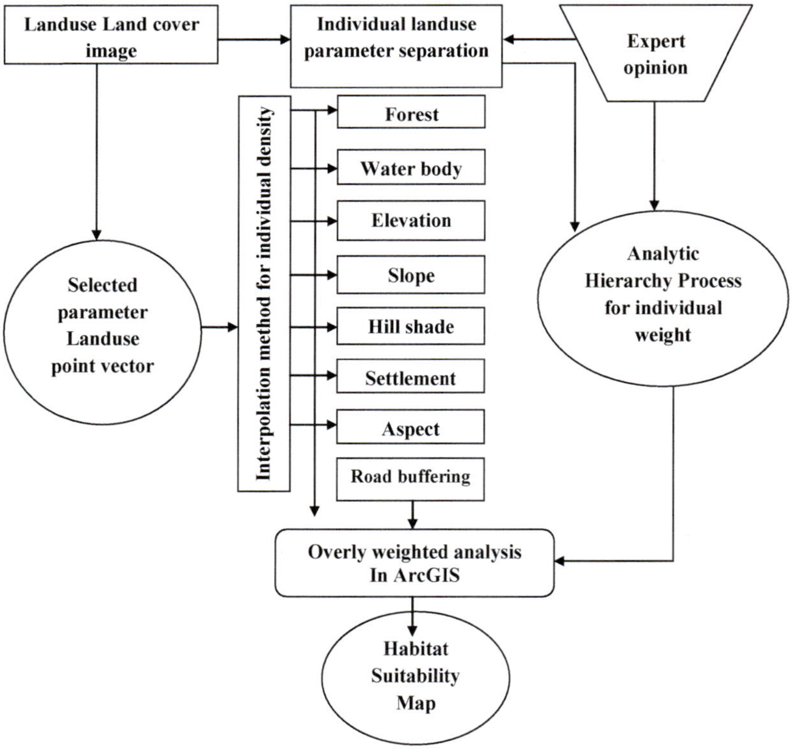

i.e., 473.59 km^2. The dense forest area (32.87%) is highly suitable. It offers better hiding place and safe shelter. Open forest (14.52%) is moderately suitable. It provides suitable sojourn ground, and other areas are (52.61%) very poorly suitable for elephant habitat (Fig. 6.17).

6.2.2.2 Water Body

Water body is very essential criterion for elephant's habitat suitability. Elephants require approximately 68.4–98.8 L (18–26 gal.) of water daily, but may drink up to 152 L (40 gal.). An adult male elephant can drink up to 212 L (55 gal.) of water in less than five minutes (Sukumar 2003). Residential elephants always stay near the permanent water source. Therefore, distance from water source is a significant factor to measure habitat quality of elephant. Entire study area is demarcated as water buffer zones considering the location of water bodies.

Buffer distance > 0.1 km is considered as high suitable area which covers 9.12% for the total habitat. Moderate suitable, low suitable and very low suitable distances from the water body buffer are 0.1–0.5 km. covering 29.50% area, 0.5–1 km. Covering 27.52% area and < 1 km covering 33.86% area, respectively (Fig. 6.18).

6.2.2.3 Elevation

Another important criterion is elevation for elephant's habitat suitability mapping. High elevation with gentle slope is suitable for elephant's habitat because generally these areas always covered with dense forest and remote from disturbances. And forest becomes more open where elevation is low. A positive relation is found between forest density and elevation. In DWS, low elevated areas are with low density off forest. Low density forest does not support better habitat qualities for elephant for safe shelter. Therefore, study considered elevation with > 454 m as high suitable area (4.30%). 454–278 m is moderate suitable area with (10.96%), 278–163 m is low suitable area with (33.68%), very low suitable < 163 m (51.07%) for the elephant habitat (Fig. 6.19).

6.2.2.4 Slope and Hill Shade

Slope is a factor that determines habitat function. Generally, high steep slope does not

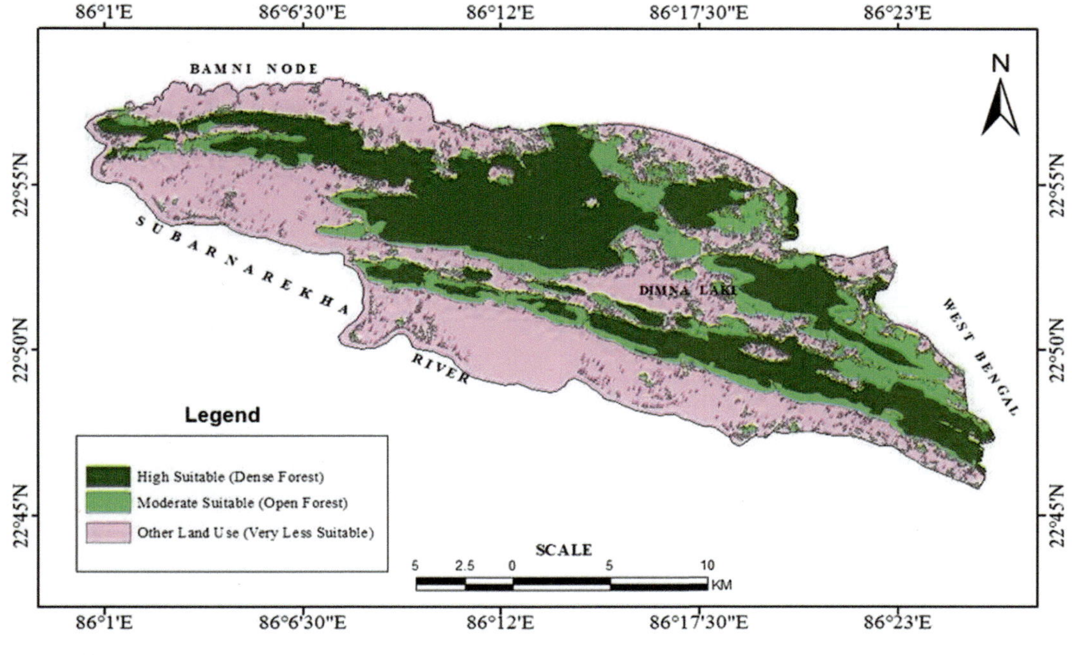

Fig. 6.17 Forest class map of DWS

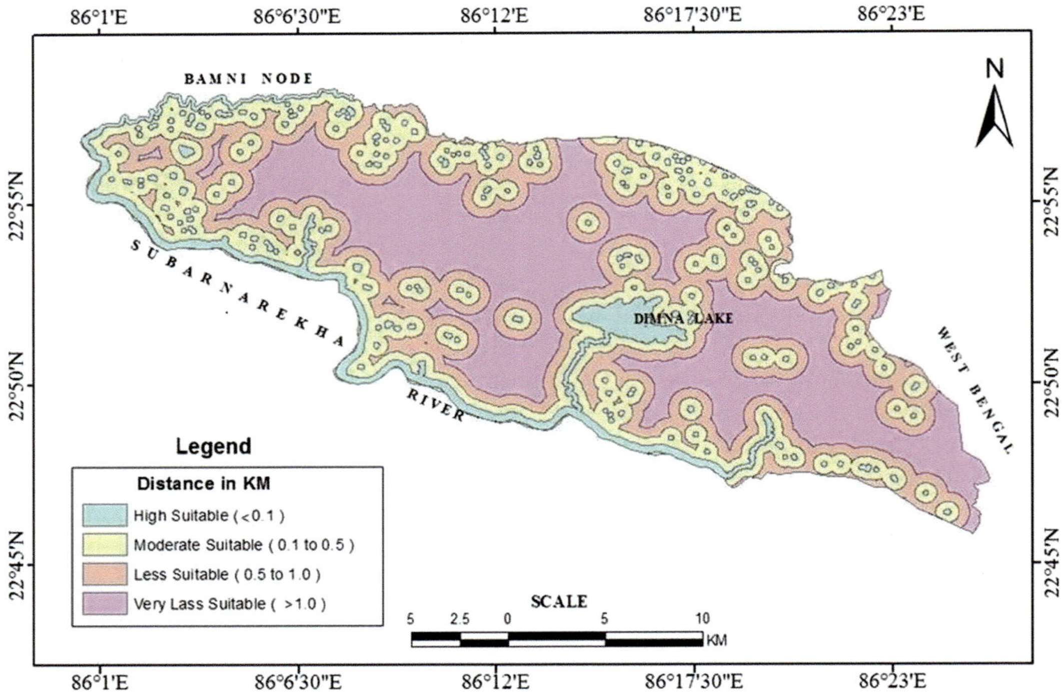

Fig. 6.18 Categorical water body buffering map of DWS

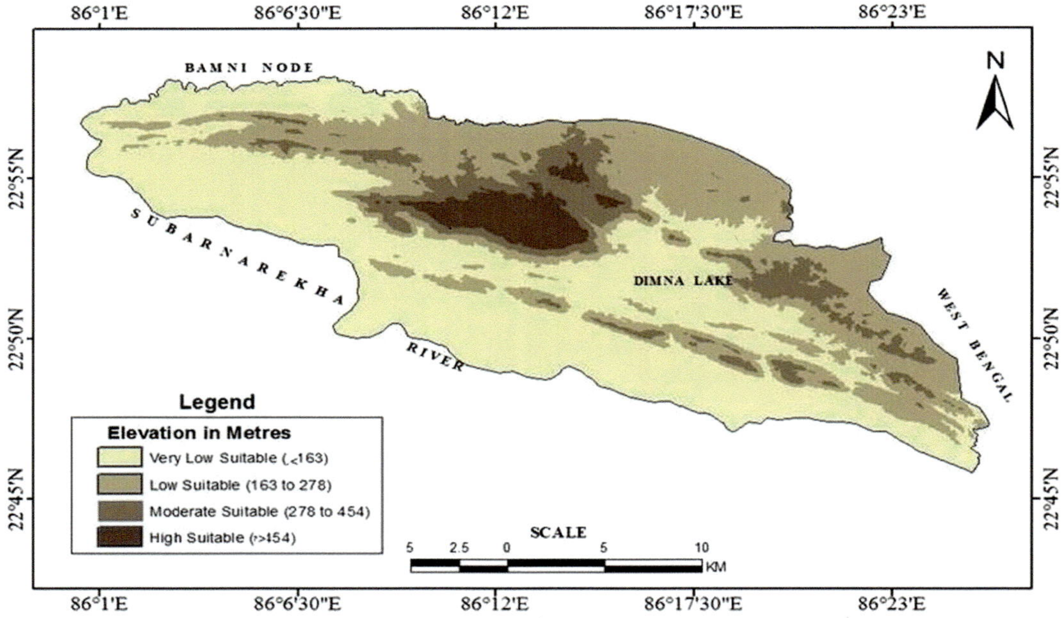

Fig. 6.19 Categorical elevation map of DWS

support as a good habitat area of animal in their home range. DWS is a hilly region. Therefore, slope becomes a factor for choosing their habitat area. Normally, large size animal prefers moderate to gentle slope areas for their resting or sheltering in the home ranges located in a plateau region. The map represents that 57.61% areas cover only 7° slope which is very highly suitable for elephant habitat. Only 8.03% areas are very poorly suitable due to high slope that above 27° (Fig. 6.20). In case of hill shade factor, only 59.71 km^2 areas is highly suitable, and 70.68 km^2 area is poorly suitable in DWS (Fig. 6.21).

6.2.2.5 Settlement

Settlement is an anthropogenic intervention that negatively influences elephant's habitat preference. In a human settled landscape, elephant's movement and colonization may be forcefully restricted which results into HEC. So the distance of settlement from forest habitat is a predicting factor that is related to elephant's habitat suitability. The study considered 0.1–0.4 km buffer distance (20% area) as least suitable for elephant's habitat because in this intermediate zone, anthropogenic intrusion become high. > 0.5 km buffer distance (38% area) is most suitable for the elephant's habitat due to less human intervention (Fig. 6.22).

6.2.2.6 Aspect

DWS is elongated from west to east direction. The orientations of the hill slop, i.e. aspect acting as a factor for forest composition and microclimate. These two factors are directly related to elephant habitat selection. Generally, northern aspect of the hill slope covered highest forest composition in north hemisphere for that reason northern aspect is highly suitable for elephant. It has been found that from the aspect map northern aspect area is 18.79% of the total area of DWS (Fig. 6.23). It will be the highly suitable area and 30.43% area is covering western face will be poorly suitable for elephant due to low quality of forest composition.

6.2.2.7 Road

Road is the most important criterion for wildlife habitat selection (Forman 2012). Road

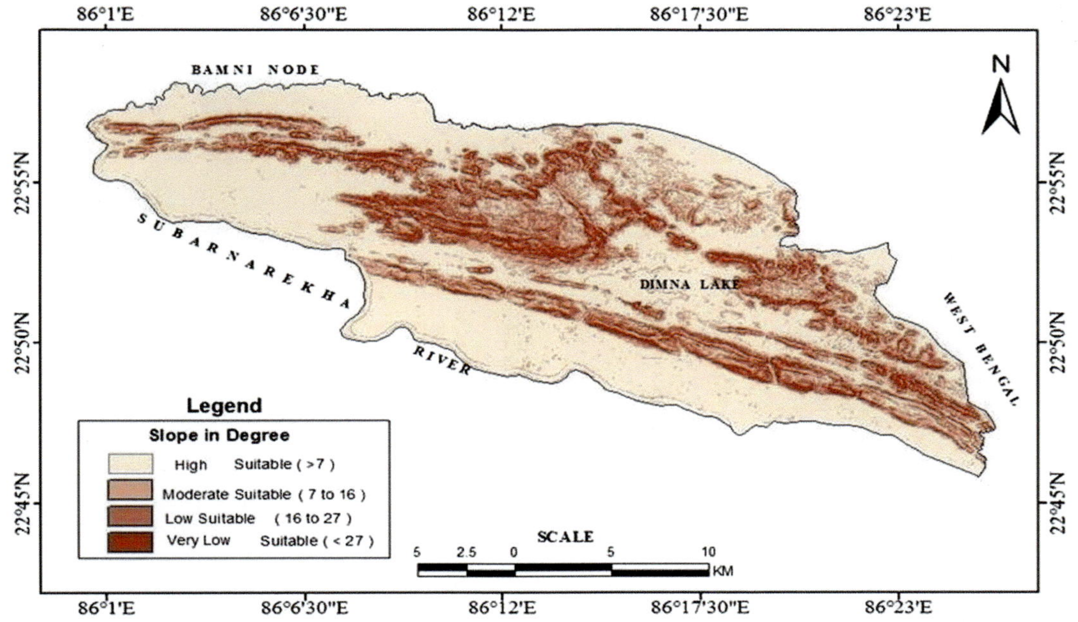

Fig. 6.20 Categorical slope distribution map of DWS

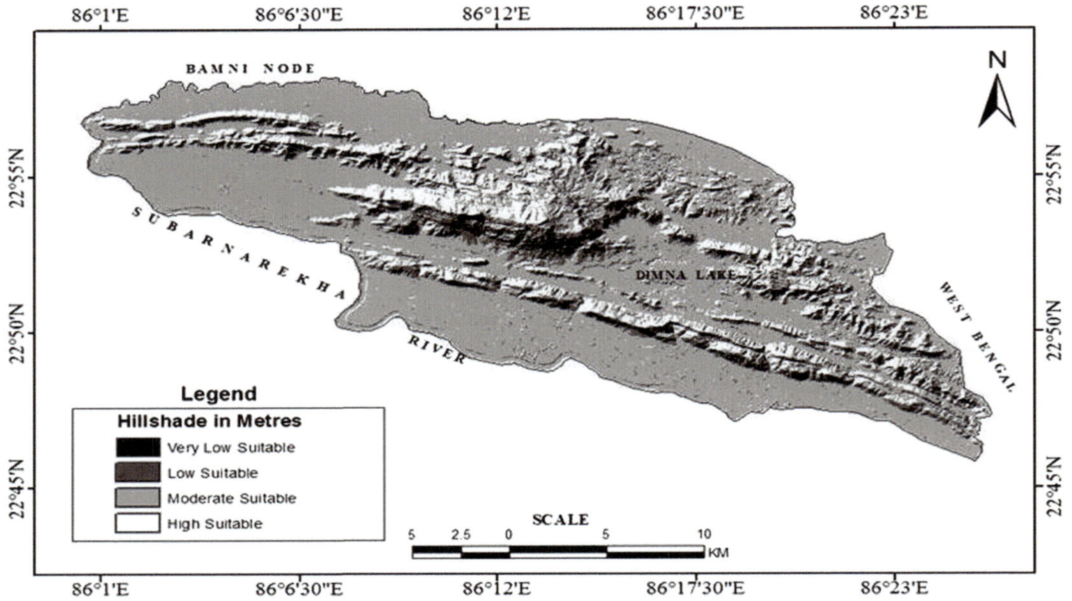

Fig. 6.21 Hill shade density map of DWS

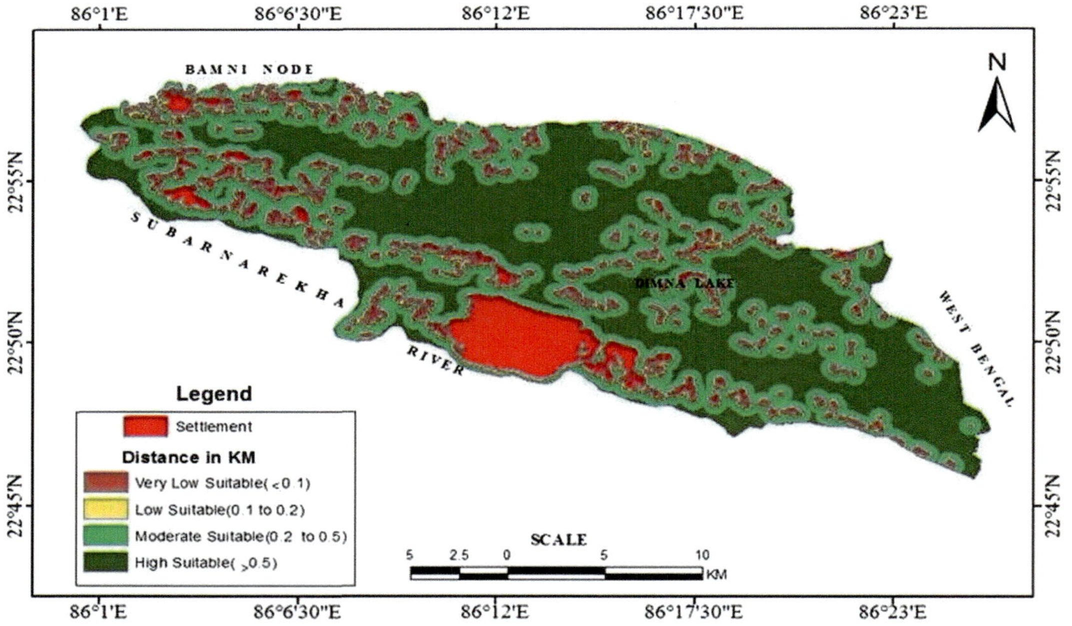

Fig. 6.22 Settlement buffering map of DWS

disturbed wildlife, so, area surrounding the road is not suitable for elephant's habitat and more distance from the road are suitable for elephant's habitat. The present work tries to buffer the entire road in four distinct zones. Habitat suitability will be increased with

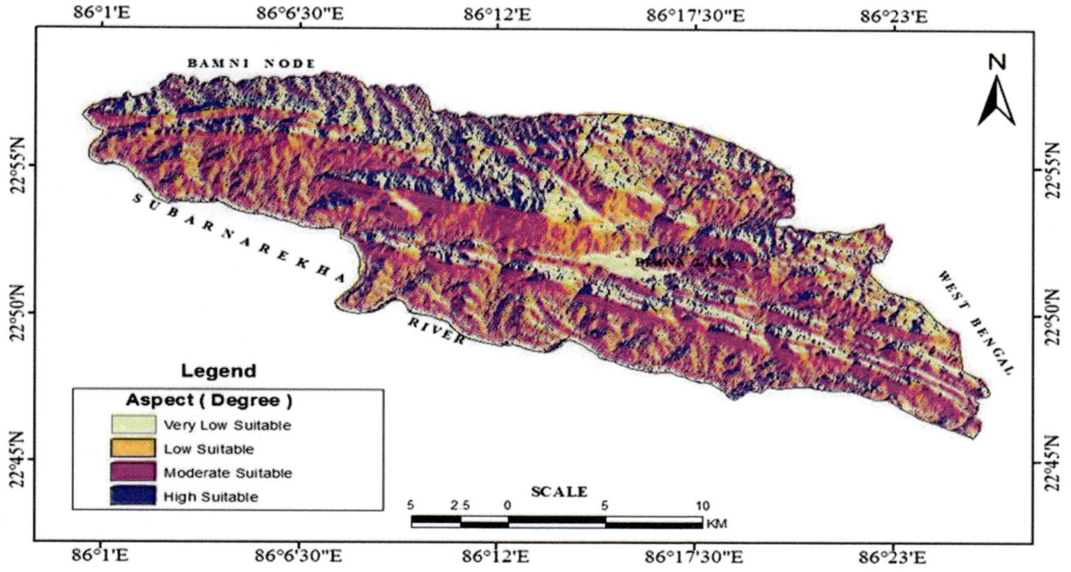

Fig. 6.23 Aspect ratio map of DWS

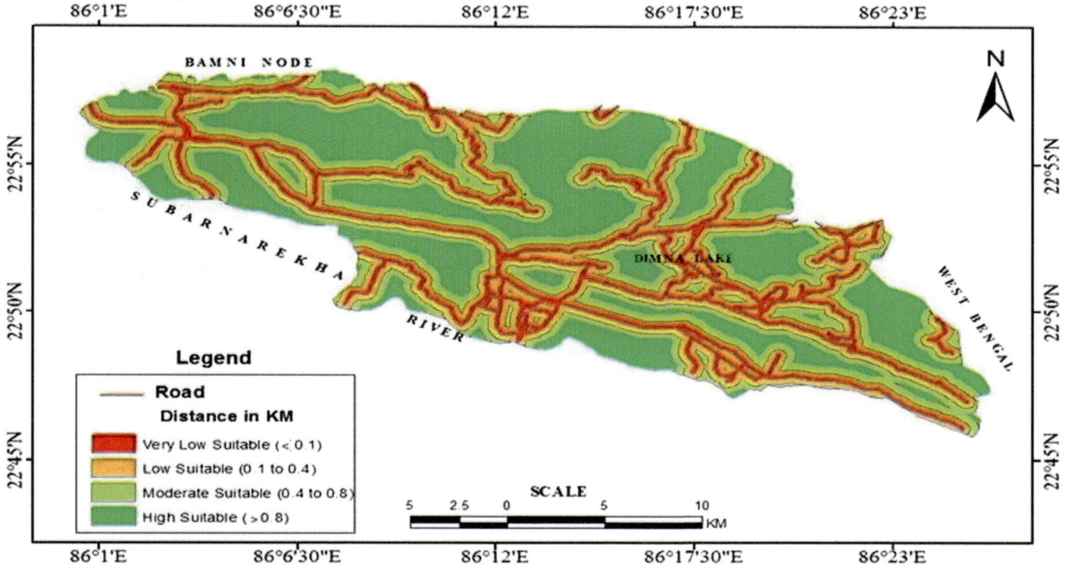

Fig. 6.24 Road buffering map of DWS

increasing buffer distance due to less significant disturbances. In this study area, < 0.1 km buffer distance (12% area) is very low suitable, 0.1–0.4 km (28%) areas low suitable, 0.4–0.8 km (27%) area moderate suitable and > 0.8 (33%) km areas are high suitable for elephants habitat (Fig. 6.24).

6.2.3 AHP Weights for Wildlife Habitat Suitability Mapping

The present research uses AHP method for habitat quality assessment. To find out the individual factors weight on habitat suitability zoning, a nine point measurement scale, i.e. 1—Equal importance, 3—Moderate importance, 5—Strong importance, 7—Very strong importance, 9—Extreme importance are used. The other values of 2, 4, 6 and 8 constitute intermediate values between two adjacent judgements. AHP is based on the assumption that some factors are more important than others in a particular circumstance (Dyer and Forman 1992). Input set factors in the raster format each input raster, is multiplied with a corresponding factor weight and then summed. The weights obtained from the AHP used in weight overlay and sum in Arc Map 10.1softwere to generate final suitability maps. It will be divided into three natural classes of suitability level.

To examine the rationality with pair-wise comparison method of AHP, it is necessary to determine the degree of consistency that has been used in developing the judgements. In AHP, an index of consistency, known as the consistency ratio (CR), is used to indicate the probability that the matrix judgements were randomly generated.

$$\mathrm{CR} = \mathrm{CI}/\mathrm{RI}, \quad (6.1)$$

where RI is the average of the resulting consistency index depending on the order of the matrix given by Saaty and consistency index (CI) is defined as

$$\mathrm{CI} = (\lambda\mathrm{max} - n)/n{-}1, \quad (6.2)$$

where λmax—n is a principal Eigen value of matrix and n is the number of our input criteria participated.

The consistency ratio (CR) is designed in such a way that if CR < 0.10, the ratio indicates a reasonable level of consistency in pair-wise comparisons; if, CR $\geq$ 0.10, the values of the ratio are indicative of inconsistent judgements (Tables 6.5 and 6.6).

6.2.3.1 Assigning Criterion Weights

All the criteria of this work assign weights using pair-wise comparison technique (Saaty 1980). These criterion weights are calculated using Microsoft excel (Table 6.5). The weighted values of any criteria depend on the user preference.

6.2.4 Overly Weighted Analysis

The habitat requirements of the elephants and the variables related to these requirements are established based on the existing information and field expert opinion. Eight major factors (forest, water body, elevation, slope, hill shade, settlement, aspect and road) are chosen as the most important determining parameters for assessing the elephant habitat suitability under DWS.

This classified individual factor map has equal interval in considerable subcriteria as a class.

Table 6.5 Weighted value for different criterion derived from pair-wise comparison technique

Criteria	Waterbody	Forest	Settlement	Road	Elevation	Hill shade	Slope	Aspect	Rank	% weight
Determining the relative criteria weights										
Waterbody	1	1	3	2	1	1	1	1	11	14.64
Forest	1	1	1	1	4	6	1	1	16	21.34
Settlement	0.33	1	1	3	1	1	1	1	9.33	12.44
Road	0.5	1	0.33	1	1	1	1	1	6.83	9.11
Elevation	1	0.25	1	1	1	1	3	1	9.25	12.33
Hill shade	1	0.17	1	1	1	1	2	1	8.17	10.89
Slope	1	1	1	1	0.33	0.5	1	2	7.33	9.77
Aspect	1	1	1	1	1	1	0.5	1	7.05	9.40
SUM	6.83	6.42	9.33	11	10.33	12.5	10.5	9	74.96	100

Table 6.6 Calculation of pair-wise comparison technique

Estimation of the consistency ratio		
The value for 'λ' is simply the average value of the consistency vector		
λ = {sum of all vector consistency/no of criteria (n)}		
λ =	9.05032	
Computation the consistency index (CI)		
CI = (($\lambda - n$)/($n - 1$)	CI =	0.15005
Consistency ratio (CR)		
CR = CI/RI	Where 'RI' is random index, depends on the no of element being compared	
CR =	0.1064157	

Subcriteria of parameters are also scored within the rank 1–4 against each class (Table 6.7) excluding forest class. In this scoring of subcriteria's high score, i.e. 4 is highly suitable for elephant habitat. Prepared raster map of 8 parameters are overlaid using the weighted sum overlay analysis tool in ArcGIS 10.1 version software. Parameter weights and subparameter individual scores finally portrays elephant habitat suitability.

6.2.5 Results and Discussions

Wildlife movement and habitat colonization are the most impotent process related to wildlife conflict as well as HEC in different parts of the World. These processes directly concerned with habitat qualities. Therefore, Habitat suitability mapping is now very essential tool for investigating different biogeographical processes (Calabrese et al. 2017). Depending on this, the present study prepare an elephant habitat suitability map (Fig. 6.25) of DWS. This map will help to address HEC characters in the study area. The result presents that 16.61 km^2 area is highly suitable for elephants habitat, 48.26 km^2 is moderately suitable and 408.72 km^2 (86.30%) area is very poorly suitable for elephant habitat in DWS. Most of the highly suitable areas cover upper portion of the hill where forest density also high. Foot hill areas are under moderate and less suitable zone where forest patches are scatter and settlement unit are doted. The villages like Khokro, Bandhdih Kaira, Banta, Tulin in DWS are very closer to highly suitable habitat areas of elephant. But it is found that HEC is very few in these villages. During the time of field investigation it is seen that, these villages are much close to dense forest and located beside the core areas of the forest. The dense forest is highly preferred by elephant for their daily biological interaction (Mandal et al. 2015). It is found that elephants do not stay long time in open area; they requires close forest for protection, resting and sleeping (Songer et al. 2012).

Generally, it has been observed from several elephant domains that HEC is frequent at the time of their movement and feeding behaviour (Roever et al. 2012). Only for food and drink, elephants move regularly from habitat to nearest habitat (Bist 2006; Williams et al. 2008). Highly suitable habitat areas support them for hide and safe shelter from any kind of disturbance (Sundaram et al. 2003). Moderate and less suitable habitat areas are the source of their food and drink. Therefore, they move to these places frequently. Asan bani, Kumari, Haludbani, mirjadih, Kadamjhor, Tankocha, Kuiani, Gerua, Punsa, Kataluka, Amdapahari these villages are closer to moderate habitat suitability zone (Fig. 6.26). HEC is a common phenomena in moderately and poorly suitable zones because elephants are coming from highly suitable areas to these areas in search of agricultural

Table 6.7 Parameters class and its score in overly weighted analysis

Selected parameters	Weight in %	Parameter sub criteria classes		Score
Water body buffering	14.64	< 0.1 0.1–0.5 0.5–1.0 > 1	km	4 3 2 1
Forest class	21.34	Dense Mixed Open	–	3 2 1
Settlement buffering	12.44	< 0.5 0.5–0.2 0.2–0.1 > 0.1	km	4 3 2 1
Road buffering	9.11	< 0.8 0.4–0.8 0.1–0.4 > 0.1	km	4 3 2 1
Elevation	12.33	< 454 454–278 278–163 > 163	m	4 3 2 1
Hill shade	10.89	High Medium Low Very low	Density	1 2 3 4
Slope	9.77	< 27 27–16 16–7 > 7	Degree	1 2 3 4
Aspect	9.40	North South East West	Ratio	4 3 2 1

crop. Agricultural crop, i.e. secondary vegetation is highly preferred by elephant for their daily food habits (Bremset Hansen et al. 2009; Kumar et al. 2010). For that reason, the moderately and poorly suitable habitat areas are facing maximum HEC than highly suitable areas. HEC like crop damage, asset loss, human death or injury, are very common incident that are found in these region frequently. Human death and injury records given by Mango Forest Range presents maximum conflict have been occurred at less suitable habitat areas (Fig. 6.27). Similar results received from field information and sanctuary reports (Table 6.8). Human death and injury are more frequent in these villages located in moderate and less suitable areas in DWS.

High and moderate suitable habitats are found in the middle portion of DWS. Especially in high suitable areas elephant stay and rest safely without any disturb. The peripheral areas are accounted as a poorly suitable habitat for elephant. But in the peripheral region, more villages are found and practicing agricultural activities throughout the year (Plate 6.1). This secondary vegetation attracts elephant from suitable habitat and caused HEC in these villages. Water inefficiency in the highly suitable areas is the reason for low HEC in these areas. Elephant moves towards the water point to

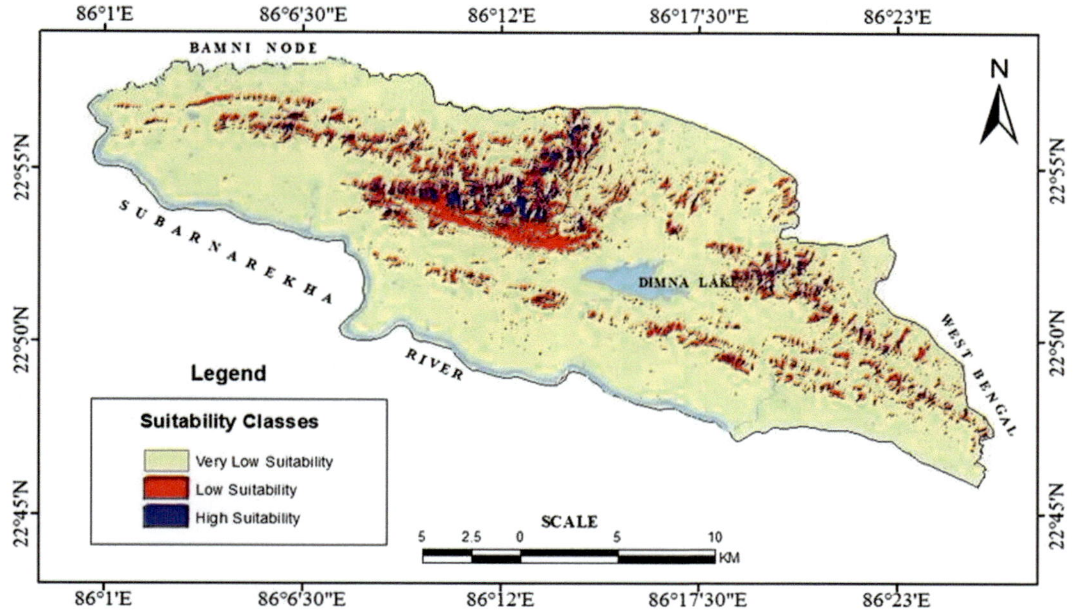

Fig. 6.25 Elephant habitat suitability map on the basis of selected criteria in DWS

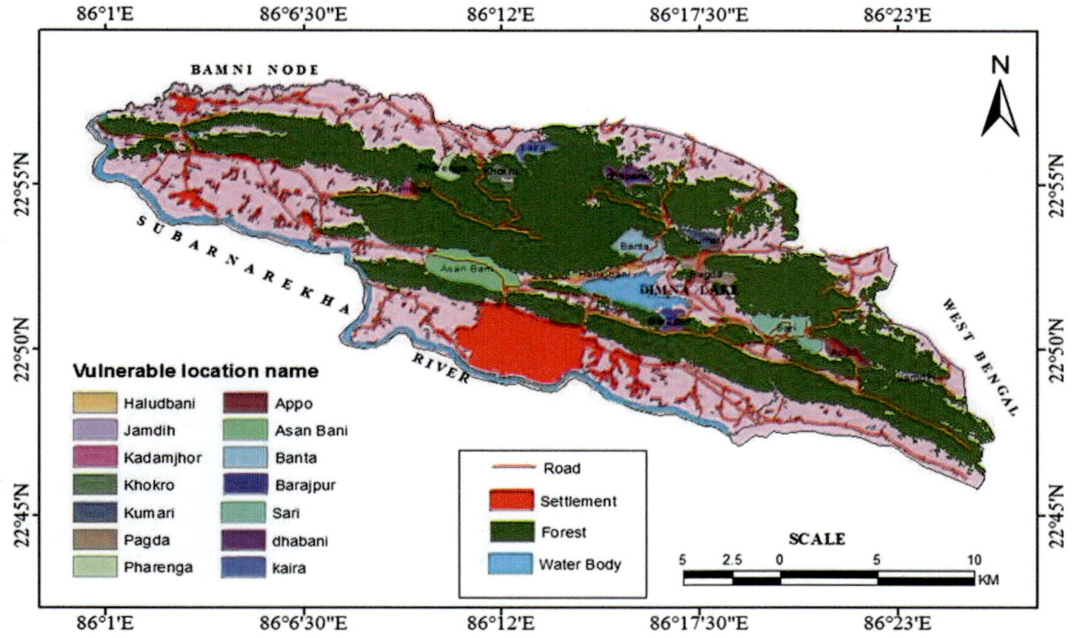

Fig. 6.26 Vulnerable villages for frequent HEC under DWS

meet their daily requirements. Available water source point in lower portion of Dalma hill, i.e. in poorly suitable areas make possibilities for coming elephants for drinking water. According to the forest staff, several human deaths were occurred during morning or evening when elephants are thirsty or busy to drink water.

Table 6.8 Persons killed by wild elephant in last 10 years in DWS

S. No.	Date	Death person's name	Village name
1	03.05.08	Nobin Singh	Ghorabonga
2	22.09.09	Romi Singh	Detakhoda
3	10.07.10	Mintonjon Mahata	Kudlonga
4	13.06.10	Dukhu Pahariya	Bagdiho
5	07.07.10	Mangol Singh	Goborghushi
6	10.08.12	Gurucharan Sabar	Goborghushi
7	21.04.12	Balaram Besra	Baruduho
8	19.07.12	Mongol Sabar	Koira
9	11.06.12	Bahadur Singh	Ghorabonga
10	07.08.12	Badal Singh	Appo
11	07.05.13	Josaf Lokra	Kudmakachar
12	24.06.13	Dheru Sabor	Goborghushi
13	14.03.13	Ramsharub Kushbaha	Garua
14	26.06.14	Ajit Sabor	Pogda
15	13.02.15	Ghasiram Murmu	Gasoa
16	05.05.15	Nimai Soran	Kongdarbara
17	16.05.15	Avi Paradiha	Asan Bani
18	25.08.15	Madhu Singh	Kadamsor
19	21.02.16	Sukdeb Bhuya	Jamboni
20	27.02.16	Saraswati Gop	Pardiho
21	10.03.16	Gopal Singh	Bagdiho
22	06.03.16	Dhishu Sabar(Valu)	Dusra
23	07.04.16	Balobir Singh	Mijadiho
24	27.01.18	Chatan Mahali	Asan Bani

Source Data collection of Mango forest office, Jharkhand

Plate 6.1 Photos of paddy cultivation in the lower portion of Dalma Hill

6.2.6 Conclusion

We have attempted the methodology for implementing eight important factors for elephant habitats suitability at DWS such as forest, water body, elevation, slope, hill shade, settlement, aspect and road. Integration of GIS and Remote Sensing are helpful in identification, assessments, and mapping the suitable habitats for elephants (Ahmad et al. 2018) by combine decision support methodology, visualization and analysis, which should considerably help in finding best suitable

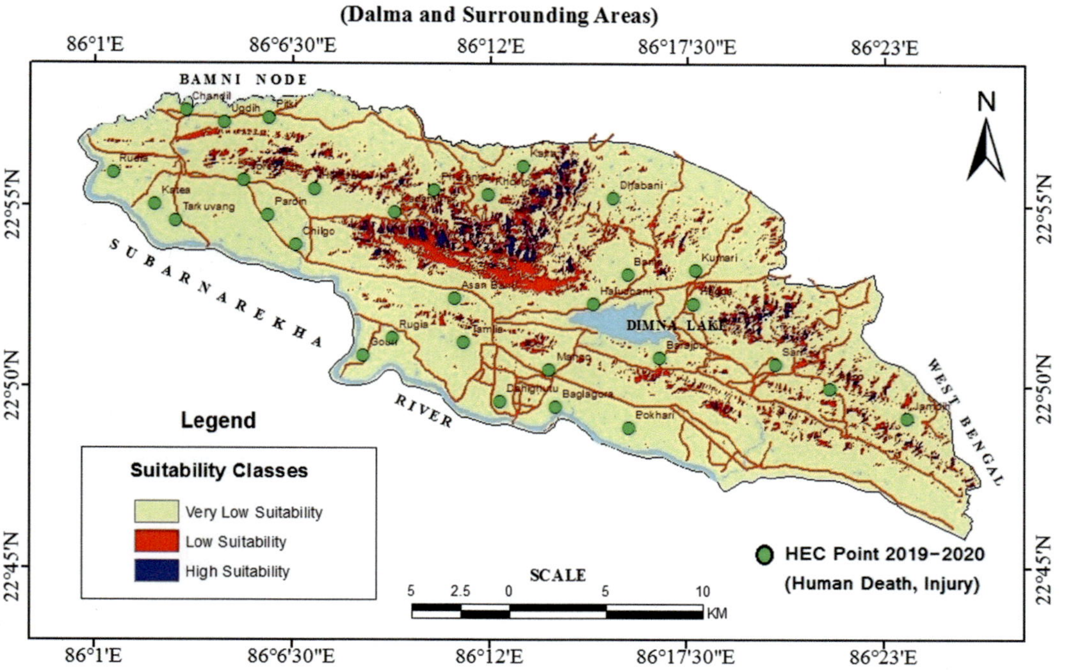

Fig. 6.27 Presents location map showing HEC in DWS (human death and injury). Conflict is high in the less suitable habitat areas. *Source* Mango Forest Range record, Jharkhand, 2019–2020

elephant's habitats at DWS in Jharkhand. After addressing the HEC condition a trend is found that elephant uses highly suitable areas for their resting only and they uses less or moderate suitable areas for sojourn ground in DWS. A senior forest officer Avishek Kumar said that, elephant regularly comes from upper portion of the hill (under highly suitable habitat areas) to lower portion (less or moderate suitable areas) for foraging and drinking. He appreciated with this result that HEC is high in this lower portion of Dalma hill. Next question is how it will be managed? Effective management of habitat corridors selection, forest patch structure and compositional quality enhancement (Mandal and Chatterjee 2018), forest core development (Mandal and Chatterjee 2019), scientific landuse alteration (Mandal et al. 2020a) water source point development in scarcity zone, fodder plant species plantation in forest core in highly suitable areas may restrict elephant movement from this region to the surrounding states. The prepared map will help to identify the habitat suitability class in DWS. Frequent awareness camp and program would be a long way in ensuring symbiotic relationship between elephants and human in the study area.

6.3 Overall Outcomes

From the conclusion part of the two case studies, it is clear that landuse land cover characteristics play an important role for wildlife habitation and demarcating their suitable habitat in a spatial context. Space-animal interaction has an influence on regional ecological circumstances that has been discussed in two case studies from two different regions. So wildlife management as well as conflict measurements both are important when we get the overall spatial characters. Not only that their function with the behaviour of wild animal should be a matter of fact. Suitable management strategies steps may be taken to improve biodiversity and manage HWC by practicable rational landuse alteration and scientific management proposals that deals with stockholders perceptions and participation in the regional ecological circumstances.

References

Ahmad F, Goparaju L, Qayum A (2018) Wild life habitat suitability and conservation hotspot mapping: remote sensing and GIS based decision support system. AIMS Geosci 4(1):66–87

Akcakaya AR (2000) Conservation and management for multiple species: integrating field research and modelling into management decisions. Environ Manage 26:S75–S83

Akinci H, Özalp AY, Turgut B (2013) Agricultural land use suitability analysis using GIS and AHP technique. Comput Electron Agric 97:71–78

Anthony B, Avery G (2008) Elephant management: a scientific assessment for South Africa. NYU Press

Areendran G, Raj K, Mazumdar S, Munsi M, Govil H, Sen KP (2011) Geospatial modeling to assess elephant habitat suitability and corridors in northern Chhattisgarh, India. Trop Ecol 52(3):275–283

Baskaran N, Varma S, Sar CK, Sukumar R (2011) Current status of Asian Elephants in India. Gajah 35:47–54

Bist SS (2006) Elephant conservation in India. An overview. Gajah 25:27–37

Bogucki DJ, Bormann FH, Box EO, Bratton SP, Dolan R, Dunn CP, Knight DH (2012) Landscape heterogeneity and disturbance, vol 64. Springer Science & Business Media

Bremset Hansen B, Herfindal I, Aanes R, Sæther BE, Henriksen S (2009) Functional response in habitat selection and the tradeoffs between foraging niche components in a large herbivore. Oikos 118(6):859–872

Burrough PA, Mcdonnell R (1998) Principles of geographical information systems. Oxford University Press, Oxford

Calabrese A, Calabrese JM, Songer M, Wegmann M, Hedges S, Rose R, Leimgruber P (2017) Conservation status of Asian elephants: the influence of habitat and governance. Biodivers Conserv 26(9):2067–2081

Chakraborty K, Mondal J (2013) Perceptions and patterns of human–elephant conflict at Barjora block of Bankura district in West Bengal, India: insights for mitigation and management. Environ Dev Sustain 15 (2):547–565

Chanda S (1996) Man-elephant conflict in South West Bengal. Joint Director of Sundarban Biosphere Reserve, West Bengal) IFS (1996 batch), 1–8

Chatterjee ND (2016) Man-elephant conflict: a case study from forests in West Bengal, India. Springer International Publishing, Switzerland

Chatterjee ND, Chatterjee S (2014) Changing habitat and elephant migration from Dalma Wildlife Sanctuary, Jharkhand to Panchet Forest Division, Bankura, West Bengal: a biogeographical analysis. In Climate change and biodiversity. Springer, Tokyo, pp 209–222

Chowdhury S, Khalid MA, Roy M, Singh AK, Singh RR (1997) Management of elephant populations in West Bengal for mitigating man-elephant conflicts. Wildlife Institute of India, Dehradun

Couvillion BR (2005) Spatial heterogeneity in forested landscapes: an examination of forest fragmentation and suburban sprawl in the Florida Parishes of Louisiana

Cushman SA (2006) Effects of habitat loss and fragmentation on amphibians: a review and prospectus. Biol Cons 128(2):231–240

Cushman SA, Chase M, Griffin C (2010) Mapping landscape resistance to identify corridors and barriers for elephant movement in southern Africa. In: Spatial complexity, informatics, and wildlife conservation. Springer, Tokyo, pp 349–367

Das S, Jeganathan C, Saptarshi M, Richa S (2018) Geospatial modelling of human-elephant Conflicts in Dalma wildlife sanctuary and its surroundings in India. Ed, Michael O'Neal Campbell in Geomatics and Conservation Biology, 77. Nova Science Publishers, New York

Datye HS, Bhagwat AM (1995) Home range of elephants in fragmented habitats of central India. J Bombay Nat Hist Soc 92(1):1–10

Desai A (2008) Notes from the Co-chairs IUCN/SSC Asian Elephant Specialist Group. Gajah 28:3

Desai AA, Hedges S (2010) Notes from the Co-chairs IUCN/SSC Asian Elephant Specialist Group. Gajah 33:3–6

Desai AA, Riddle HS (2015) Human-elephant conflict in Asia. Supported by: US Fish and Wildlife Service Asian Elephant Support

Dickman AJ (2010) Complexities of conflict: the importance of considering social factors for effectively resolving human–wildlife conflict. Anim Conserv 13 (5):458–466

Dramstad W, Olson JD, Forman RT (1996) Landscape ecology principles in landscape architecture and land-use planning. Island Press

Dyer RF, Forman EH (1992) Group decision support with the analytic hierarchy process. Decis Support Syst 8 (2):99–124

Estes JG, Othman N, Ismail S, Ancrenaz M, Goossens B, Ambu LN, Palmiotto PA (2012) Quantity and configuration of available elephant habitat and related conservation concerns in the Lower Kinabatangan floodplain of Sabah, Malaysia. PloS One 7(10)

Fahrig L, Arroyo-Rodríguez V, Bennett JR, Boucher-Lalonde V, Cazetta E, Currie DJ, Eigenbrod F, Ford AT, Harrison SP, Jaeger JA, Koper N (2019) Is habitat fragmentation bad for biodiversity? Biol Cons 230:179–186

Farina A (2008) Principles and methods in landscape ecology: towards a science of the landscape, vol 3. Springer Science & Business Media

Fernando P, Kumar MA, Williams AC, Wikramanayake E, Aziz T, Singh SM (2008) Review of human-elephant conflict mitigation measures practiced in South Asia

Forman RTT (1995) Land mosaic: the ecology of landscape and regions. Cambridge University Press, Cambridge, England

Forman RT (2012) Safe passages: highways, wildlife, and habitat connectivity. Island Press

Forman RT, Deblinger RD (2000) The ecological road-effect zone of a Massachusetts (USA) suburban highway. Conserv Biol 14(1):36–46

Gaynor KM, Branco PS, Long RA, Gonçalves DD, Granli PK, Poole JH (2018) Effects of human settlement and roads on diel activity patterns of elephants (*Loxodonta africana*). Afr J Ecol 56 (4):872–881

Guha M (2017) Human-elephant conflict in South West Bengal–I: study on Fodder Plants of the Elephants of Dalma Herd. J Econ, Environ Soc 2(1):22–27

Hedges S, Gunaryadi D (2010) Reducing human–elephant conflict: do chillies help deter elephants from entering crop fields? Oryx 44(1):139–146

Hill CM (2018) Crop foraging, crop losses, and crop raiding. Annu Rev Anthropol 47:377–394

Holbrook JD, Squires JR, Bollenbacher B, Graham R, Olson LE, Hanvey G, Savage SL (2019) Management of forests and forest carnivores: relating landscape mosaics to habitat quality of Canada lynx at their range periphery. For Ecol Manage 437:411–425

Imam E, Tesfamichael GY (2013) Use of remote sensing, GIS and analytical hierarchy process (AHP) in wildlife habitat suitability analysis. J Mater Environ Sci 4 (3):460–467

Kanagaraj R, Araujo MB, Barman R, Davidar P, De R, Digal DK, Lamichhane BR (2019) Predicting range shifts of Asian elephants under global change. Divers Distrib 25(5):822–838

Kulandaivel S (2010) A paradigm shift in the elephant depredation in South Bengal. Divisional Forest Officer, Bankura North Division, Personal Communication

Kumar MA, Mudappa D, Raman TRS (2010) Asian elephant Elephas maximus habitat use and ranging in fragmented rainforest and plantations in the Anamalai Hills, India. Trop Conserv Sci 3(2):143–158

Kupfer JA (2012) Landscape ecology and biogeography: rethinking landscape metrics in a post-FRAGSTATS landscape. Prog Phys Geogr 36(3):400–420

Kushwaha SPS, Roy PS (2002) Geospatial technology for wildlife habitat evaluation. Trop Ecol 43:137–150

Li H, Wu J (2007). Landscape pattern analysis: key issues and challenges. In: Keytopics in landscape ecology. Cambridge University Press

Liu X, Bronsveld MC, Toxopeus AG, Kreijns MS (1997) GIS application in research of wildlife habitat change: a case study of the giant panda of Wolong nature reserve. J Chin Geogr 7(4):51–60

Loarie SR, Van Aarde RJ, Pimm SL (2009) Fences and artificial water affect African savannah elephant movement patterns. Biol Cons 142(12):3086–3098

Malczewski J (1999) GIS and multicriteria decision analysis. John Wiley and Sons, New York

Mandal M, Chatterjee ND (2018) Quantification of habitat (forest) shape complexity through geo-spatial analysis: an ecological approach in Panchet forest division in Bankura, West Bengal. Asian J Environ Ecol 6:1–8

Mandal M, Chatterjee ND (2019) Forest Core demarcation using geo-spatial techniques: a habitat management approach in Panchet Forest Division, Bankura, West Bengal, India. Asian J Geogr Res, 1–8

Mandal M, Chattarjee ND (2020a) Land use alteration strategy to improve forest landscape structural quality in Radhanagar forest range under Bankura district. Eurasian J for Sci 8(1):1–10

Mandal M, Chatterjee ND (2020b) Spatial alteration of fragmented forest landscape for improving structural quality of habitat: a case study from Radhanagar Forest Range, Bankura District, West Bengal, India. Geol Ecol Landsc, 1–8

Mandal M, Chettarjee ND, Hazra J (2015) Elephant migration and colonization in Bankura district, West Bengal, India. Vidyasagar University. Indian J Geogr Environ 14:46–52

McGarigal K, Cushman S (2005) The gradient concept of landscape structure [Chapter 12]. In: Wiens JA, Moss MR, (eds) Issues and perspectives in landscape ecology. Cambridge University Press, pp 112–119

McGarigal K, Marks BJ (1995) FRAGSTATS: spatial pattern analysis program for quantifying landscape structure. Gen. Tech. Rep. PNW-GTR-351. Portland, OR: US Department of Agriculture, Forest Service, Pacific Northwest Research Station. 122 P, 351

McIlroy JC (1978) The effects of forestry practices on wildlife in Australia: a review. Aust for 41(2):78–94

Mendas A, Delali A (2012) Integration of multicriteria decision analysis in GIS to develop land suitability for agriculture: application to durum wheat cultivation in the region of Mleta in Algeria. Comput Electron Agric 83:117–126

Miller JR (2015) Mapping attack hotspots to mitigate human–carnivore conflict: approaches and applications of spatial predation risk modeling. Biodivers Conserv 24(12):2887–2911

Mokarram M, Aminzadeh F (2010) GIS-based multicriteria land suitability evaluation using ordered weight averaging with fuzzy quantifier: a case study in Shavur Plain, Iran. Int Arch Photogrammetry, Remote Sens Spat Inf Sci 38(II):508–512

Mumby HS, Plotnik JM (2018) Taking the elephants' perspective: remembering elephant behaviour, cognition and ecology in human-elephant conflict mitigation. Front Ecol Evol 6:122

Öztürk D, Batuk F (2010) Konumsal Karar Problemlerinde Analitik Hiyerarsi Yönteminin Kullanılması. Yıldız Teknik Üniversitesi Sigma Mühendislik ve Fen Bilimleri Dergisi 28:124–137

Panja U, Mistri B (2018) Human-Elephant Conflict in Sonamukhi CD Block of Bankura District, West Bengal. Space Cult, India 5(3):106–128

Parker EG, Osborn VF, Hoare ER, Niskanen SL (2007) A training course for community-based approaches in Africa. Trainer's manual. Elephant Pepper Development Trust. Livingston, Zambia and IUCN/SSC, AfESG, Nairobi, Kenya

Pokharel SS, Singh B, Seshagiri PB, Sukumar R (2018) Lower levels of glucocorticoids in crop raiders: diet quality as a potential 'pacifier' against stress in free ranging Asian elephants in a human production habitat. Anim Conserv

Pradhan NMB, Wegge P (2007) Dry season habitat selection by a recolonizing population of Asian elephants Elephas maximus in lowland Nepal. Acta Theriol 52(2):205–214

Ranjan A, Anand A, Kumar Singh R, Sivathanu V (2016) LU/LC change detection and forest degradation analysis in Dalma Wildlife Sanctuary using 3S Technology: a case study in Jamshedpur-India. AIMS Geosci 2:273–285. https://doi.org/10.3934/geosci.2016.4.273

Reza MIH, Abdullah SA, Nor SBM, Ismail MH (2013) Integrating GIS and expert judgment in a multi-criteria analysis to map and develop a habitat suitability index: a case study of large mammals on the Malayan Peninsula. Ecol Ind 34:149–158

Rodgers WA, Panwar HS (1988) Planning a wildlife protected area network in India

Roever CL, Van Aarde RJ, Leggett K (2012) Functional responses in the habitat selection of a generalist megaherbivore, the African savannah elephant. Ecography 35(11):972–982

Ruda A, Kolejka J, Silwal T (2018) GIS-Assisted Prediction and Risk Zonation of Wildlife Attacks in the Chitwan National Park in Nepal. ISPRS Int J Geo Inf 7(9):369

Saaty TL (1977) A scaling method for priorities in hierarchical structures. J Math Psychol 15:57–68

Saaty TL (1980) The analytic hierarchy process: planning, priority setting, resource allocation. McGraw-Hill International, New York, NY, USA

Saaty TL (1990) An exposition of the AHP in reply to the paper 'remarks on the analytic hierarchy process.' Manage Sci 36:259–268

Saaty TL (1994) Fundamentals of decision making and priority theory with the AHP. RWS Publications, Pittsburgh, PA, USA

Saaty TL (2004) Mathematical methods of operations research. Dover Publications, Mineola, pp 415–447

Santra AK, Pan S, Samanta AK, Das S, Halder S (2008) Nutritional status of forage plants and their use by wild elephants in South West Bengal, India. Trop Ecol 49 (2):251

Singh AK (2006) Ecological investigation of human-elephant conflicts in South West Bengal. Saurashtra University

Singh AK, Singh RR, Chowdhury S (2002) Human-elephant conflicts in changed landscapes of south West Bengal, India. Indian for 128(10):1119–1132

Sitompul AF, Griffin CR, Rayl ND, Fuller TK (2013) Spatial and temporal habitat use of an Asian Elephant in Sumatra. Animals 3(3):670–679

Songer M, Sampson C, Williams C, Forrest J, Gyeltshen K, Huy K, Yulianto K (2012) Mapping habitat and deforestation in WWF elephant priority landscapes. Gajah 36:3–10

Sundaram B, Varma S Venkataramanand A, Sukumar R (2003) The Asian elephant (Elephas mqximus): its habitat, states and distribution in Arunachal Prodesh, India. GAJAH Journal of the Asian elephant Specialist Group

Sukumar R (1994) Wildlife-human conflict in India: an ecological and social perspective. Social ecology. Oxford University Press, New Delhi, pp 303–317

Sukumar R (2003) The living elephants: evolutionary ecology, behaviour, and conservation. Oxford University Press

Viljoen PJ (1989) Spatial distribution and movements of elephants (Loxodonta Africana) in the northern Namib Desert region of the Kaokoveld, South West Africa/Namibia. J Zool 219(1):1–19

Vogiatzakis IN (2003) GIS-based modelling and ecology: a review of tools and methods. Department of Geography, University of Reading

Von Gerhardt K, Van Niekerk A, Kidd M, Samways M, Hanks J (2014) The role of elephant Loxodonta Africana pathways as a spatial variable in crop-raiding location. Oryx 48(3):436–444

Wiersma YF, Huettmann F, Drew CA (2011) Introduction. Landscape modeling of species and their habitats: history, uncertainty, and complexity. In: Predictive species and habitat modeling in landscape ecology. Springer, New York, pp 1–6

Williams AC, Johnsingh AJ, Krausman PR, Qureshi QAMAR (2008) Ranging and habitat selection by Asian elephants (Elephas maximus) in Rajaji National Park, north-west India. J Bombay Nat Hist Soc 105:24–33

Woodroffe R, Thirgood S, Rabinowitz A (eds) (2005) People and wildlife, conflict or co-existence? (No. 9). Cambridge University Press

7 Operational Landscape Alteration Techniques to Improve Ecological Quality of Forest Habitat: Case Studies in the Fragmented Habitats

Abstract

This Chapter will deal with the management methods of Human–elephant conflict. Hence it will discuss how the Landscape management methods will be used to enhance habitat quality for wildlife conservation with special reference to elephants. Corridor demarcation, Habitat gap minimization through habitat corridor, Biodiversity Bridge and tunnel, Perennial water source development inside the habitat, as well as other biological and mechanical measures will be very useful to prepare their own model for Forest Landscape Planning and Wildlife Management.

7.1 Habitat Quality Enhancement by Landuse Planning

The objective of landuse planning is to propose sustainable management for wildlife degradation (Muruthi 2005). Loss of local endemic species is a continuous ecological process which has significant temporal causes that connected with land alteration strategy. Natural habitat modification through human intervention is common and being regularized from independence in West Bengal. As a result wildlife in Bengal gradually becomes out of order by separate segments and also fragmentized. Habitat characteristics in human animal conflict zones in South West Bengal are structurally complex in nature. The shape of the forest habitat is highly irregular and curvilinear. Generally, these patterns show high structural fragmentation. As a result the wild animals are exposed. They roam from one habitat patch to another habitat patch for better habitat condition. Land alteration is one of the systematic ways, for getting long-term ecological benefit. These factors may further responsible for habitat restriction a colonization of wild species as well as elephants in south western State of Bengal, specifically selected forest ranges under the influence zone in Bankura district (Fig. 7.1). Present study drives some case studies to understand what type of land alteration will be suitable for improvement of structural quality of habitat.

7.1.1 Habitat Structural Ecological Quality Improvement (Case Studies from Selected Forest Ranges Under Bankura District)

Habitat structural composition gives some essential ecological possibilities which control activities like species movement, animal colonization and extinction (Cushman 2006; Li and Wu 2007). Habitat structural composition is the different geometric shape of habitat patch like habitat edge, habitat core, habitat fragmentation,

M. Mandal and N. Das Chatterjee, *Geo-Spatial Analysis of Forest Landscape for Wildlife Management*, GIScience and Geo-environmental Modelling, https://doi.org/10.1007/978-3-031-33606-5_7

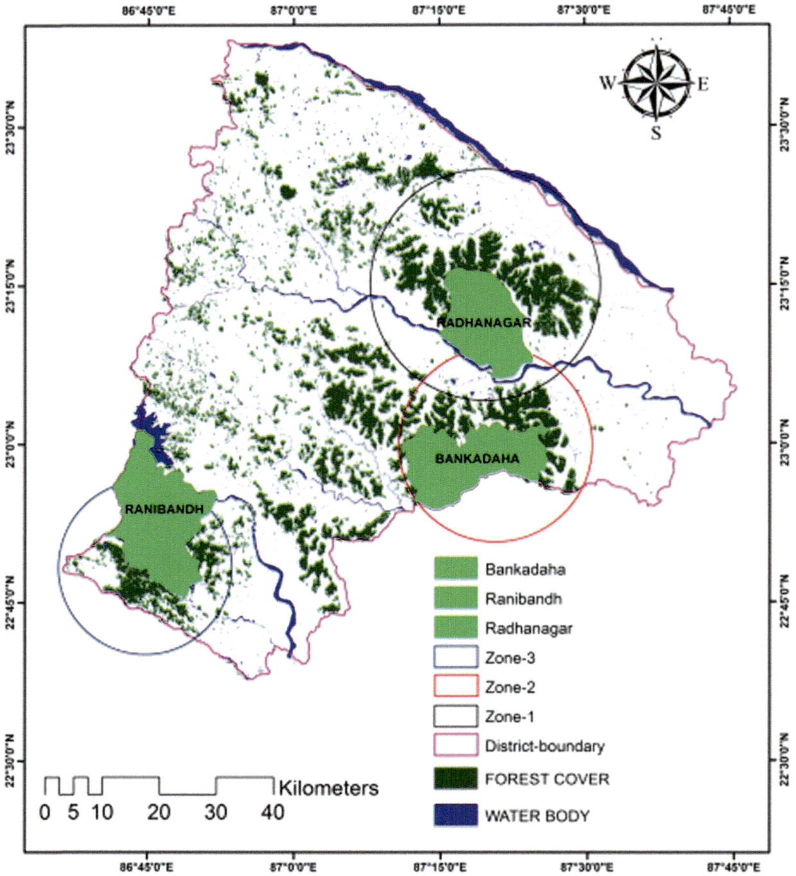

Fig. 7.1 Selected three forest ranges in different three zones under Bankura district considered for comparison analysis

connectedness and patch size, etc. Poor habitat structure refers to high fragmentation, less core area, very low effective connectivity etc. make an imbalance unsupported ecosystem (Cushman et al. 2010). In a human altered and modified landscape, systematic landscape design or planning is required to improve and manage habitat structural quality (Dramstad et al. 1996; Mandal et al. 2020a). Plantation is one of the natural ways which is common way for increasing forest habitat amount. But it does not consider structural quality of the forest. Structural management of the forest habitat should implement in strategic areas of a landscape by plantation. Therefore, selection of plantation plot or area is an essential part of forest management (Li and Wu 2007). It also amplified overall quality of the habitat. The investigation tries to shows the proper area for plantation through landuse conversion techniques. Some predicated and possible landuse conversion criteria are applied for this hypothetical management. Thus, hypothetically two type of alteration of land is considered. One is alteration of barren land into forest and second is some selected encroached area into forest land. For this analysis, alteration maps of three different forest ranges (Fig. 7.1) prepared to justify the hypothesis. After that ecological indices of altered lands are calculated and values compared with real or actual landscape to know the acceptable methods for selecting areas of plantation.

7.1.1.1 Methods

Case study proposes two types of landscape alteration methods to compare habitat quality with existing landscape (Fig. 7.2). These are barren land converted into forest land (Fig. 7.4) and some specific encroached areas (Fig. 7.3) converted to forest areas (Fig. 7.5). Generally,

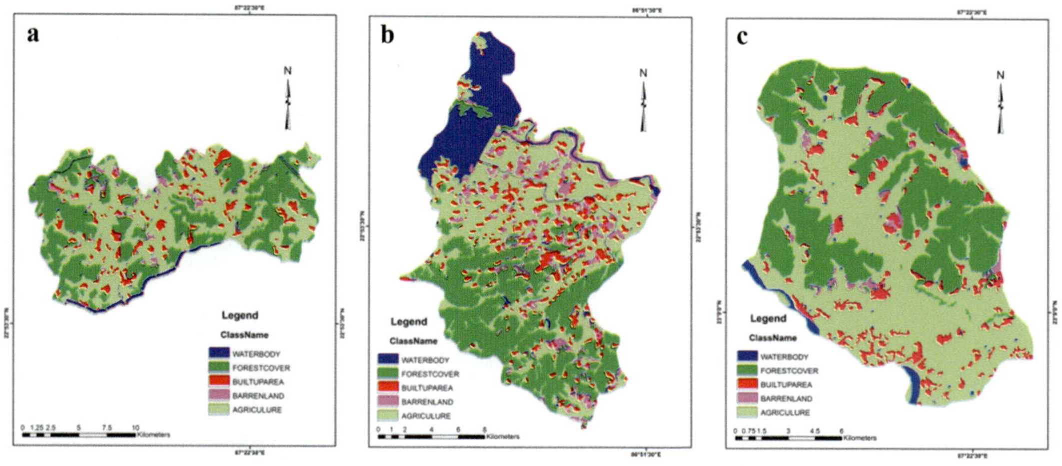

Fig. 7.2 Landuse maps of selected forest range. **a** Bankadaha, **b** Ranibandh and **c** Radhanagar

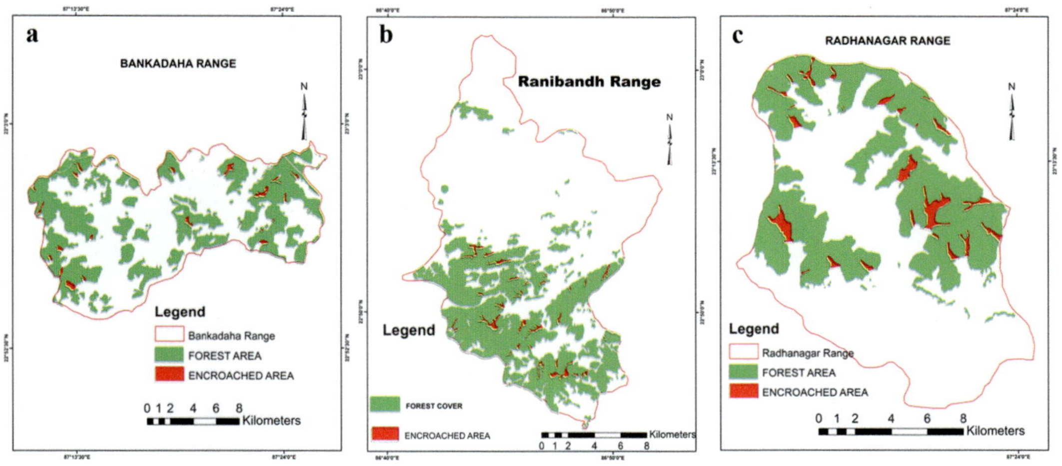

Fig. 7.3 Encroachment area (red marked area) maps of selected forest range **a** Bankadaha, **b** Ranibandh and **c** Radhanagar

forest cover areas are increased through plantation in barren land or vacant land by plantation drives. Plantation or any afforestation program will not be successful until the structural quality of the forest habitat is addressed. Habitat good like maximum forest core, less fragmented patch, minimum edge effect and less anthropogenic intrusions stated by Dramstad et al. (1996). Usually these criteria are not depending on forest cover only (Mandal et al. 2018, 2019) but all the factors are depending both for structural configurations of the patch as well as amount of forest. Thus condition wise classified categorical maps are equipped to get the actual scenario of the structural configuration of forest patches. These conditional categorical maps may be evaluated through the extension tools of ArcGIS-10.3 version. After analysing the ecological indices like TCAI, PLAND, MCA at 300 m specified edge depth, TE, ED, MSI, AWMPFD and MPS values are tabulated. Value of these indices help to understand the nature of habitat quality which is described in previous chapters. Then the tabulated indices values are compared with same values of existing landscape. The comparison may indicate, the efficient types of ecological

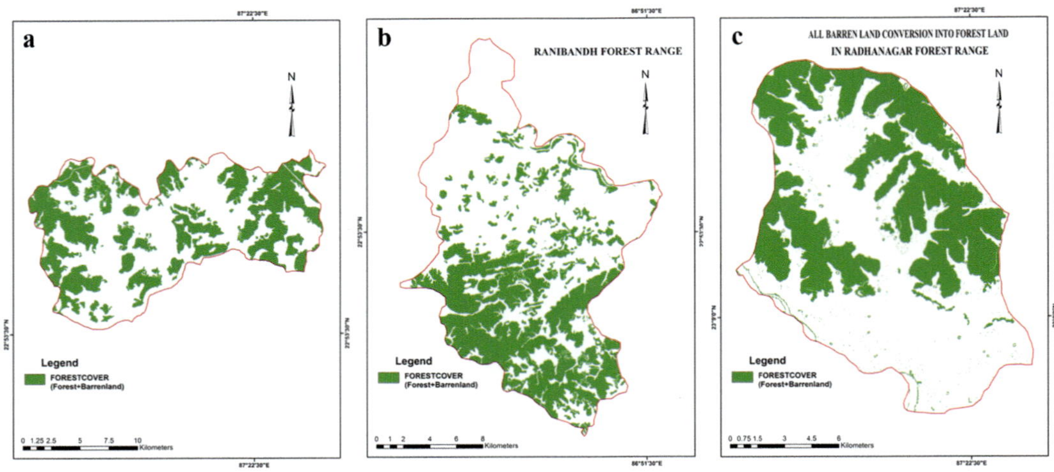

Fig. 7.4 All barren land altered into forest land conversion maps of selected three forest ranges

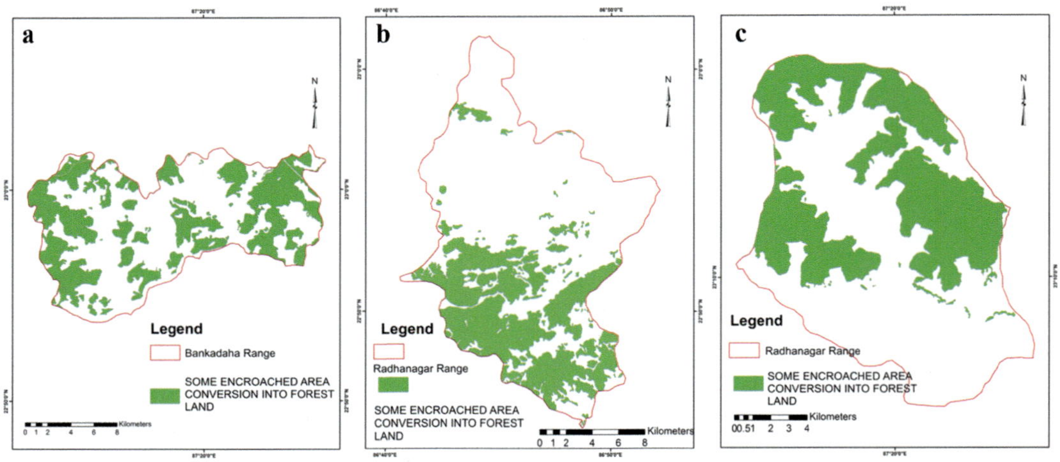

Fig. 7.5 Encroached area altered into forest land maps of selected three forest ranges

alteration suitable to improve habitat quality. Methods of all landscape ecological indices those are used for this analysis are explained in earlier chapters-4.

Ultimately, the work tries to bring out the exact place or spatial unit for plantation in fragmented forest lands. Maximum benefits will be achieved when this type of plantation method is implemented in the selected location. On the other hand this study also shows that plantation in a scattered manner in barren lands in the fragmented forest is ecologically less significant and it increases further fragmentation in the landscape as a whole.

7.1.1.2 Discussion

In discussion section land alteration conditions are named using abbreviated forms for easy understating i.e. ‘R’ represents existing landscape, ‘E’ represents encroached area into forest and ‘B’ for all barren land converted into forest land (Table 7.1).

Table 7.1 Values of different ecological indices against selected forest range against different landscape condition

Case condition	Forest Class area CA	TCAI	MCA ha	MSI	AWMPFD	TE In m	ED m/ha	MPS ha	NUMP
RANIBANDH = R	8661.75	39.42	65.66	1.59	1.22	507,000.00	58.53	120.30	72.00
RANIBANDH = B	10,854.25	32.03	54.32	1.64	1.21	849,600.00	78.27	63.48	171.00
RANIBANDH = E	8981.25	51.00	109.05	1.55	1.20	433,200.00	48.23	124.74	72.00
BANKADAHA = R	9551.75	42.87	78.75	1.76	1.15	498,200.00	52.16	156.59	61.00
BANKADAHA = B	10,138.25	40.22	76.94	1.86	1.15	588,800.00	58.08	142.79	71.00
BANKADAHA = E	9891.75	49.86	107.22	1.71	1.14	456,000.00	46.10	162.16	61.00
RADHANAGAR = R	7340.75	53.98	283.04	1.77	1.18	285,900.00	38.95	319.16	23.00
RADHANAGAR = B	7718.25	51.07	197.09	1.74	1.19	352,200.00	45.63	241.20	32.00
RADHANAGAR = E	7971.00	68.51	682.59	1.61	1.13	215,500.00	27.04	346.57	23.00

From the tabulated values it is found that forest area increased in both condition E and B from existing landscape, i.e. R. Highest amount of forest area found in B condition which is 2192.25 ha for Ranibandh, 586.5 ha for Bankadaha and 443.5 ha for Radhanagar than R condition, i.e. existing landscape. In case of E condition, i.e. encroached area into forest, forest area increased but low in amount than B condition (Fig. 7.6a). It is interesting to observe that in case of E condition, amount of MPS has improved more than R and B condition in three forest ranges (Fig. 7.6b). This result shows that encroached area is better for plantation which will straighten the forest habitat over the landscape.

In case of shape complexity, E alteration values declined than R and B alteration i.e. habitat quality improved in case of E condition. Edge density values dropped in case of E alteration in these three forest ranges than R and B condition which is better indication of E (Fig. 7.7b). It will minimize habitat edge effect. Similar result is also found in AWMPFD and MSI with declining habitat structural fragmentation (Figs. 7.7a and 7.8). Therefore encroached area conversion is significant for strategic plantation area to decrease edge contrast.

Habitat dependence is also another important factor. For habitat dependence, MCA and TCAI values improved in E alteration for three forests ranges (Fig. 7.9a, b). It is interesting to know that forest habitat area is raised in case of B condition than R and E. But habitat dependency (Core index value) declines due to more shape complexity in three forest ranges. The results suggest that structure of the patch is important than patch area.

The comparison presents that forest plantation should be programmed in encroached forest area to improve habitat quality. A few amount of land if planted in encroached areas it may give maximum ecological benefit. Habitat structural quality enhancement is possible without habitat loss (Fahrig 2003; Mandal et al. 2020b) when potential encroached area inside the habitat converted into forest cover (Fig. 7.10). Forest plantation should be implemented in those identified places it will increase forest structural quality (Mandal et al. 2021a). By this process, forest core area will also be increased. Increased core of the forest offers better quality for endemic animal species like elephants and also other wildlife. They can stay safely and extend their movement between fragmented habitat patches. As a result of it, edge effect will be decreased due to decline of edge density. Same method may be fruitful in several forests of West Bengal where habitat fragmentation is an urgent issue.

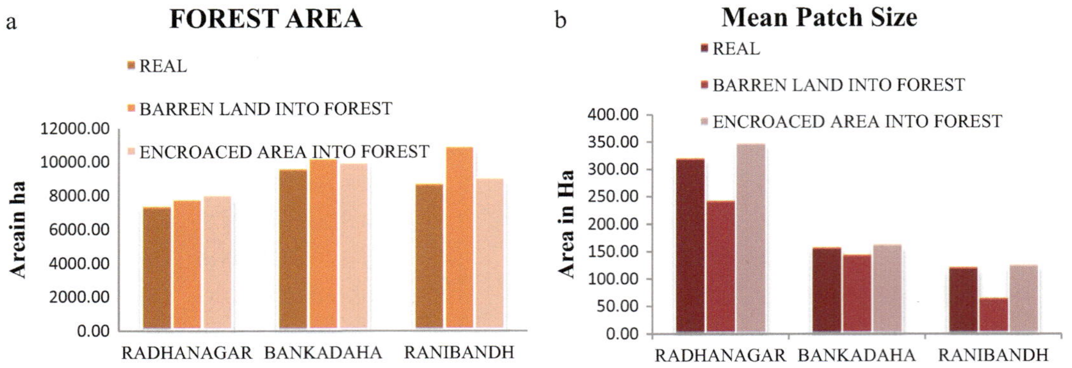

Fig. 7.6 Represents the index value deviation of Forest Class Area (FCA) (**a**) and Mean Patch Size (MPS) of B and E in respect of R (**b**)

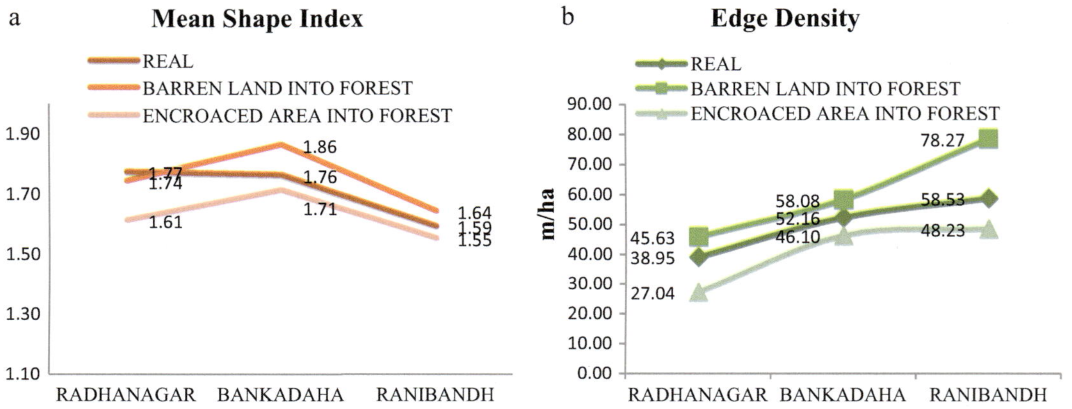

Fig. 7.7 Represents the index value deviation of MSI of E and B landscape condition in respect of R condition (**a**), ED (**b**)

Fig. 7.8 Represents the index value deviation of AWMPFD of E and B landscape condition in respect of R condition

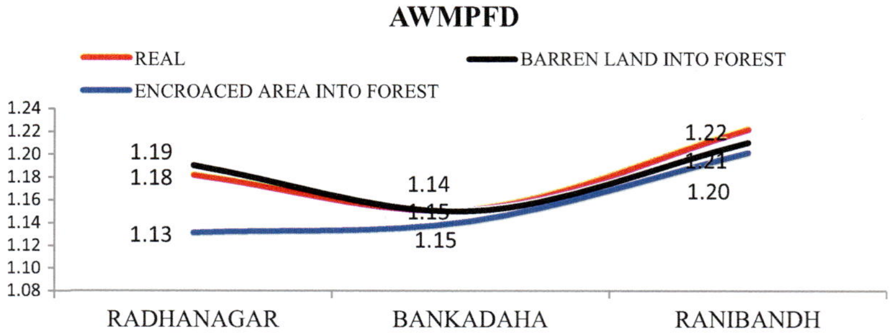

Technically, encroached forest area is converted into forest land by forest shape modification after demarcating the encroached areas. In this chapter we presented some demarcated encroached areas in different influence zone (Fig. 7.11) under Bankura district. When this modified encroached areas will be planted, habitat quality will be better. This method may be beneficial using least amount of forest plantation and may be a long-term sustainable management.

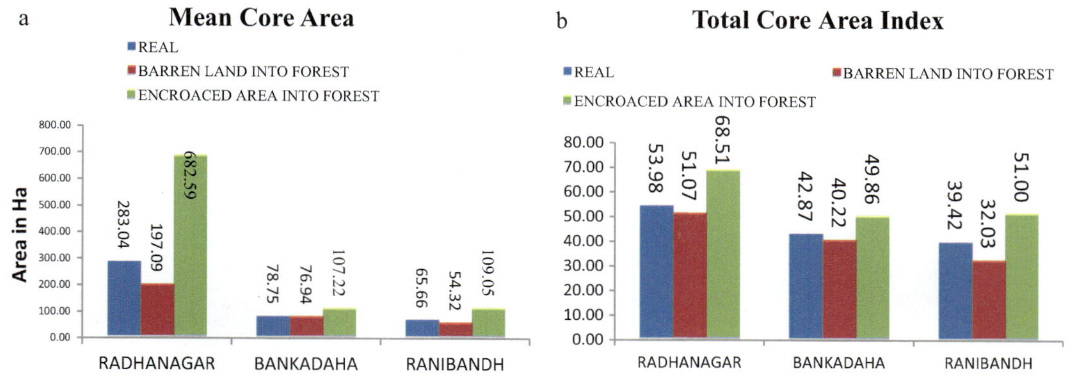

Fig. 7.9 Represents the index value deviation of MCA (**a**) and TCAI of E and B landscape condition with respect to R condition (**b**)

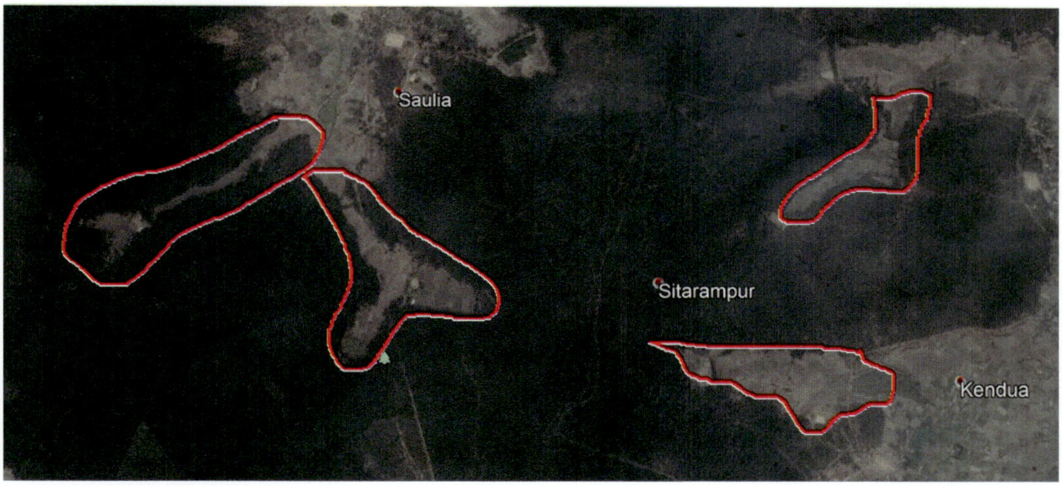

Fig. 7.10 Forest encroachment areas (red polygon area) in Sitarampur muza under Pirargari forest range. *Source* Google image 2019

7.1.2 Scientific Way of Plantation

Selection of the plant species and selected area to be planted are two major important objectives for any forest regeneration program (Mouquet et al. 2002). This present study suggests that way of plantation should consider two things. First, the selection of existing local plant species which have a strong capability to improve maximum canopy, undergrowth and layering of the forest (selected plant species in Table 7.2). Second is the selection of plant species which have productive food supply capacity as fodder for wildlife species (Harper 1977) for herbivores animal inside the forest.

7.1.2.1 Selected Plant Species for Habitat Quality Enhancement

We suggest some selected local plant species which have strong influence on ecosystem and quality of the habitat. These plant species improve habitat conditions that may restrict the frequent movement of migratory wild animal (Chatterjee 2016; Mandal et al. 2021b).

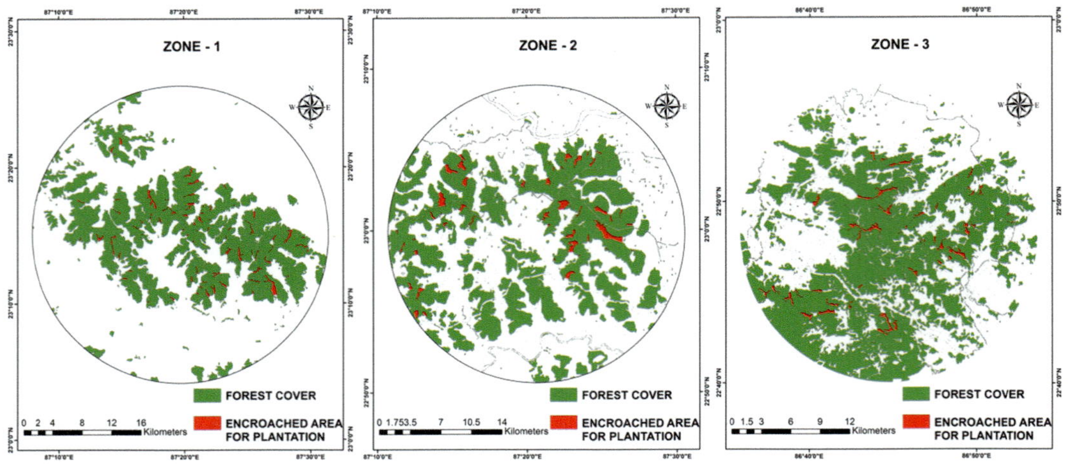

Fig. 7.11 Encroached area demarcation maps of zone-1 in Bankura North Forest Division, zone-2 in Panchet Forest Division and zone-3 in Bankura South Forest Division

Table 7.2 Scientific name of plant species which are selected for plantation for increasing habitat quality of elephants in three identified zones

Local name	Scientific name
Mahwa	*Madhuca indica*
Bandar lathi/Amaltas	*Cassia fstula*
Peasal	*Pterocarpus marsupium*
Sal	*Shorea robusta*
Kend	*Diospyros melanoxylon*
Kusum	*Schleichera trijuga*
Asan	*Terminalia tomentosa*
Rahara	*Soyamida febrifuga*
Dhaw	*Anogeissus latifolia*
Bahera	*T. belerica*
Banana	*Musa acuminate*
Jackfruit	*Artocarpus integrifolia*
Karchmola (Geol)	*Lannea grandis, v*

Deep green portion of the right side map, i.e. Joypur forest patch under zone-2 influence areas (Fig. 7.12) if considered as plantation area with selected plant species then it may provide more ecological benefit in future. This method is not only effective to reduce structural fragmentation of habitat and edge effect but it may provide a better hiding place for wild animal (Desai and Hedges 2010; Desai 1991).

7.1.3 Connectivity Development

In case of wildlife connectivity, two major consideration, i.e. habitat gap and habitat corridor are very important landscape component to improve safe movement (Forman 1995; Sawyer et al. 2011). Elephants usually follow a specific migration route within their entire home range in a fragmented forest landscape (Osipova et al.

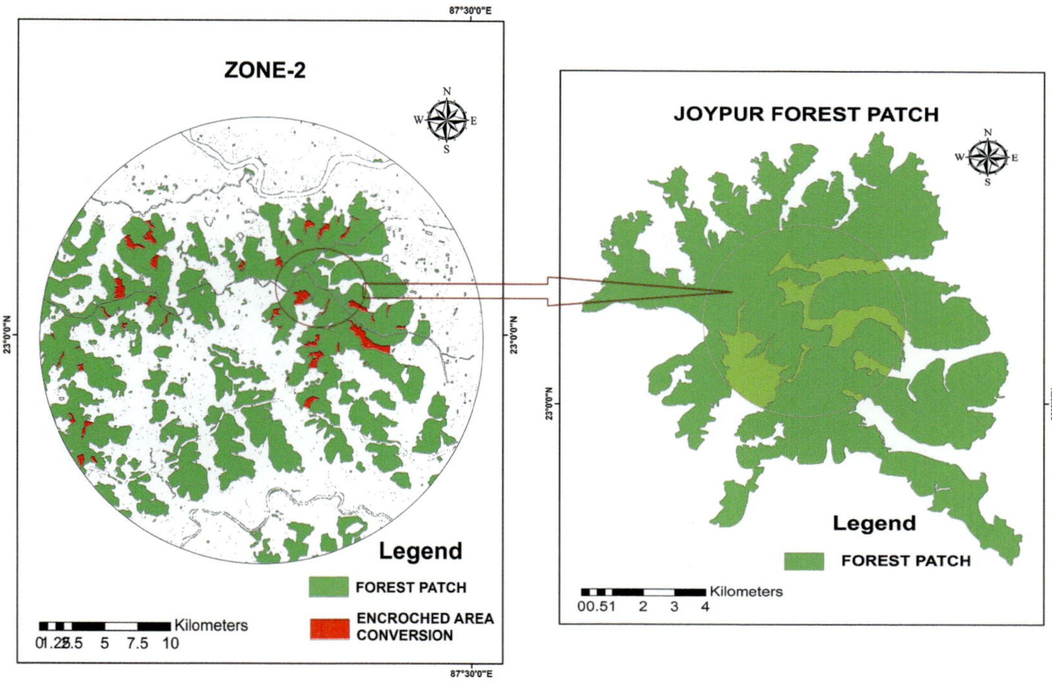

Fig. 7.12 Exact plantation area of adjoining Joypur forest, in influence zone-2 for wild life habitat conservation

2019). If habitat gap is more and continuous, then elephant's movement may be rerouted or haphazard (Fernando et al. 2008; Tripathy et al. 2021). It is a vital fact for occurrence of frequent HEC.

7.1.3.1 Corridor Demarcation

Identification and demarcation of animal corridors is an important drive in biodiversity conservation (Farina 2006, 2008) especially for controlling the movement of elephant and manage HEC (Abhijitha et al. 2021). For that reason present study tries to demarcate the possible elephant movement route for the district by the help of different information. Technical methods of corridor demarcation were explained in detail in the previous chapter-5. Another simple method y also be applied for sketching actual route of elephant in this district. For that purpose, forest range wise elephant route information recollected from different sources like locality survey, beat office information and previous literatures. Then, a proposed route map is prepared (Fig. 7.13) with the help of temporal information. This assessment was done by two conjugative years survey information (2018–2019). After demarcating the movement route it has been found that the common movement route passes through zone-1 and zone-2. Rare movement routes start from these two zones towards north-west where forest patch is more fragmented and isolated. These rare movements are found during winter season. Elephants only touch zone-3 at the time of entrance from Paschim Medinipur district.

7.1.3.2 Habitat Gap Minimization Through Habitat Corridor Development

The movement route in these three influence zones under this district do not follow the forest tract but it extends over the forest gap also (Mandal et al. 2021a). Habitat gap due to isolation and fragmentation are two major factors for wildlife conservation (Fernando et al. 2008). These gaps usually characterized by settlement areas or agricultural lands. Hence these areas may be the most conflict prone zones. Stepping stone and hedgerows would be proposed landscape unit for habitat corridor development, i.e.

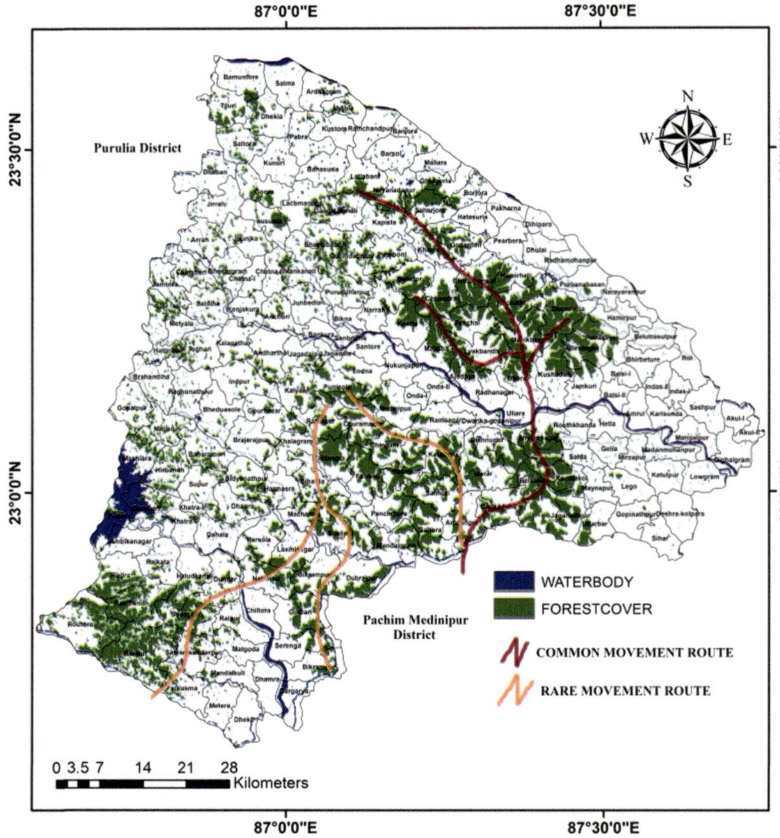

Fig. 7.13 Common and rare migration routes of elephants under Bankura district

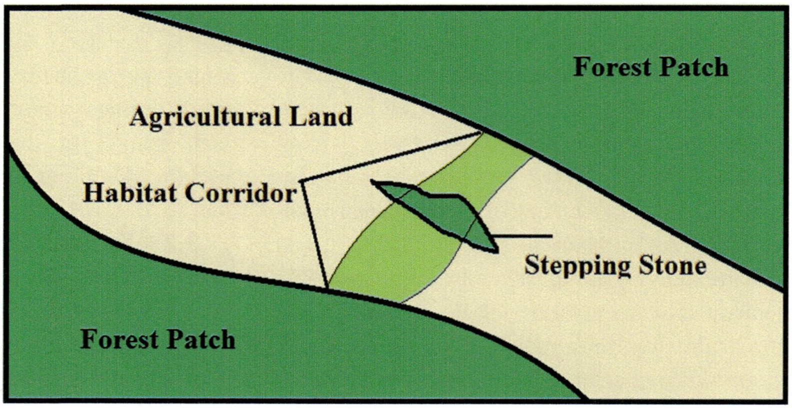

Fig. 7.14 Habitat corridor design and development through hedgerow and stepping stone by plantation to link between two nearest forest habitat patches

habitat gap minimization between nearest forest patches or to link two adjacent forest patches (Fig. 7.14). HEC may be reduced if habitat corridors develop following the wildlife migration route (Baldus et al. 2007; Naidoo et al. 2018).

Local plant species should be used for corridor development. This method brings stability in ecosystem as well as improves wildlife biodiversity (Cushman 2006). More and frequent habitat linkages offer better habitat connectivity.

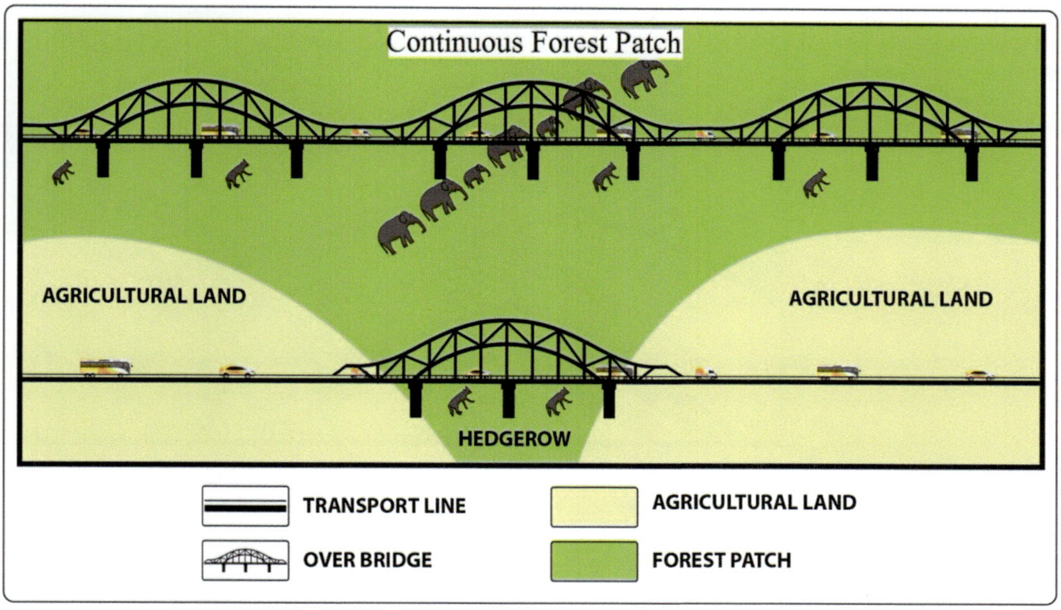

Fig. 7.15 Model to reduce wild animal collision through human corridors in between adjoining forest patches

7.1.3.3 Biodiversity Bridge and Tunnel

Biodiversity Bridge is a new concept in wildlife conservation (Forman 2014; Geijzendorffer et al. 2016; Kraus et al. 2021; Fan and Lindshield 2022). Kansabati canal and several transport network lines are the main hindrences for free wildlife as well as elephant movement. Wildlife risk is maximum in zone-2 where Kharagpur-Adra railway line (Peardoba), NH-60 (Bankadaha) and SH-2 (Joypur) roads are passing through different elephant corridors. Kansabati canal is passing across all forest patches under this district and create a barrier for wildlife easy movement. In zone-2, Biodiversity Bridge and biodiversity tunnel are immediate requirement to protect wildlife from disturbances and road accidents (Fig. 7.15). Elephants are crossing through NH-60, SH-2 and railway track near Peardoba and Joypur forest and they become victim of road accident. In these areas, Biodiversity Tunnel is appropriate landscape modification for safe connectivity and conservation (Fig. 7.16). Kansabati canal is concretized in the both sides therefore, elephant could not easily cross it (Mandal et al. 2015; Chatterjee 2016). For this reason, elephants are forced to move towards forest fringe areas for foraging (Mandal et al. 2021c). Biodiversity Bridge is an architectural model if it is implemented over Kansabati canal then that may help elephant's to move from one forest patch to nearest forest patch safely and easily in zone-2 (Fig. 7.17) especially in Basudevpur, Chagulia, Marar and Upper Peardoba forest region.

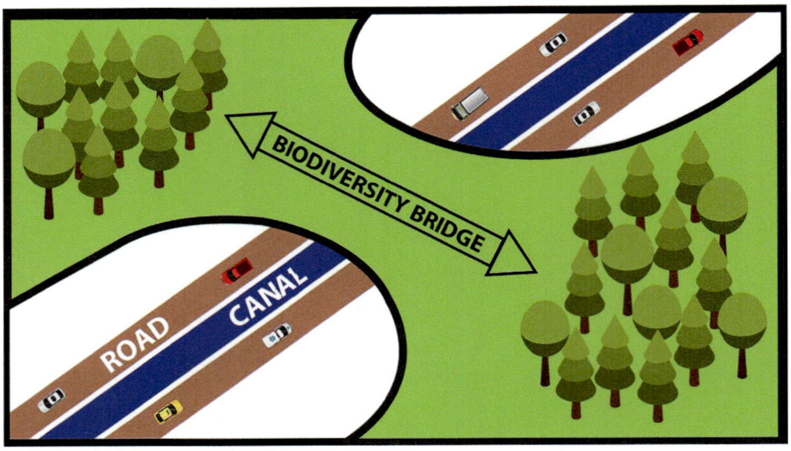

Fig. 7.16 Biodiversity Bridge model for safe movement of wild animal between nearest forest patches

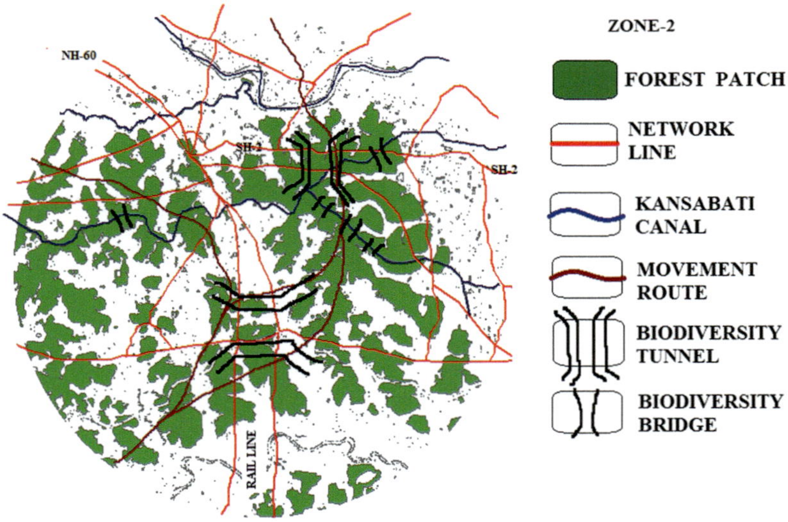

Fig. 7.17 Possible location for Biodiversity Bridge and Tunnel establishment in zone-2 for safe and free animal movement

Box 7.1: Indian Golden Jackal road accidents became normal in many parts of West Bengal. Especially winter to post winter season these incidents become frequent. These pictures are taken from a District High-Way from Patashpur to Debra, West Bengal

If no death is expected, then why, when the Jackal (*Canis aureus*) die?

Every day two to three bodies are lying on the road in my way of school. I don't know what to do or what not to do to save them. They have as much right as we have in the world. There is no one to think about them. But they always benefit people for balancing ecosystem.

7.1.4 Perennial Water Source Development Inside the Habitat

The improved elephant habitat may be those habitats where food and water are abundant throughout the year (Sukumar 1989; Fernando et al. 2008; De Knegt et al. 2011; Neupane et al. 2019). In Zone-3, i.e. the part of Mayur Jharna Elephant Reserve elephant's appearance are rare though habitat structural quality and connectedness both are better (Mandal et al. 2021c). Insufficient water and food sources are responsible mainly during dry season. In this zone-3 water points are insufficient in number and availability (Fig. 7.18). Elephants usually come from the forest core area to the fringe area to drink water after sunset and before sunrise. Artificially preserved water sources near the human settlements like canal, ponds, tank, etc. attract elephants during dry season (Thouless 1996). This is one of the main causes of HEC in South West Bengal. Water source point within the forest should be developed to restrict elephants inside the forest. Not only source point but perennial water supply throughout the year is a crucial fact. If water points are constructed inside the forest and reserve during dry season then they usually not come outside of the forest (Mandal et al. 2021d). When elephants do not prefer to go outside of the forest and like to stay into the forest the probability of HEC will be less.

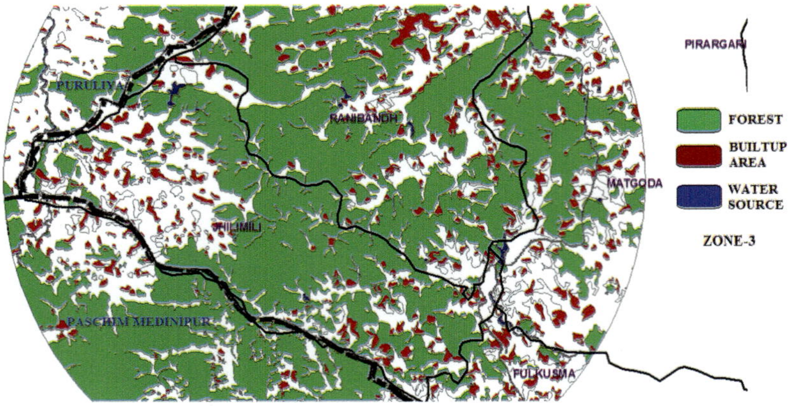

Fig. 7.18 Represents minimum water source points in zone-3 region with forest cover and built-up area

Box 7.2: Death of elephants due to hit by train in North Bengal forest and electrification in crop field in South Bengal

World Elephant Day was first observed on 12th August 2012. The special day was stated to observe by two Canadian film directors and the Reintroduction Foundation of Thailand. The purpose of observing the day is to generate more awareness on elephant conservation. Statistics from different sources show that the numbers of elephant deaths are increasing in the world. The main reason for occurance of such events are the conflict between humans and elephants.

In North Bengal elephant hit by train frequently in 2004 due to broad gauge line from Siliguri to Alipurduar junction. So the number of deaths of elephant has increased. The railway from Siliguri to Alipurduar was upgraded from meter-gauge to broad gauge in 2004. After the alteration, the number of elephants killed by trains increased. Fifty-three elephants died from 1973 to 2013. Of these, 30 elephants were killed by trains from 2004 to 2013. Elephant deaths due to train collisions decreased in 2015–2016 but continue to increase since 2017. Six elephants have been killed by trains in this region. Elephants are dying due to poison or electronic shock in the agricultural fields.

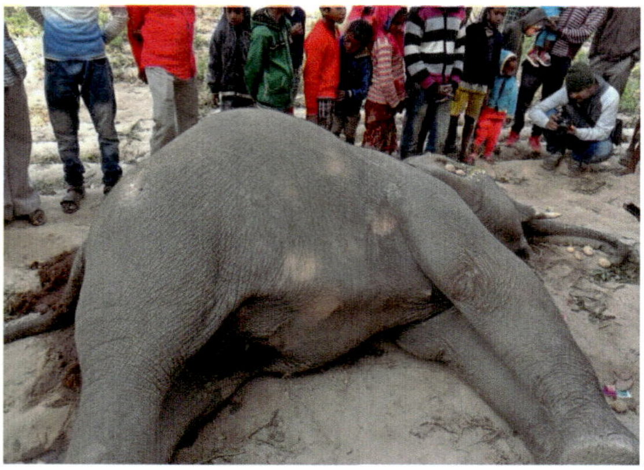

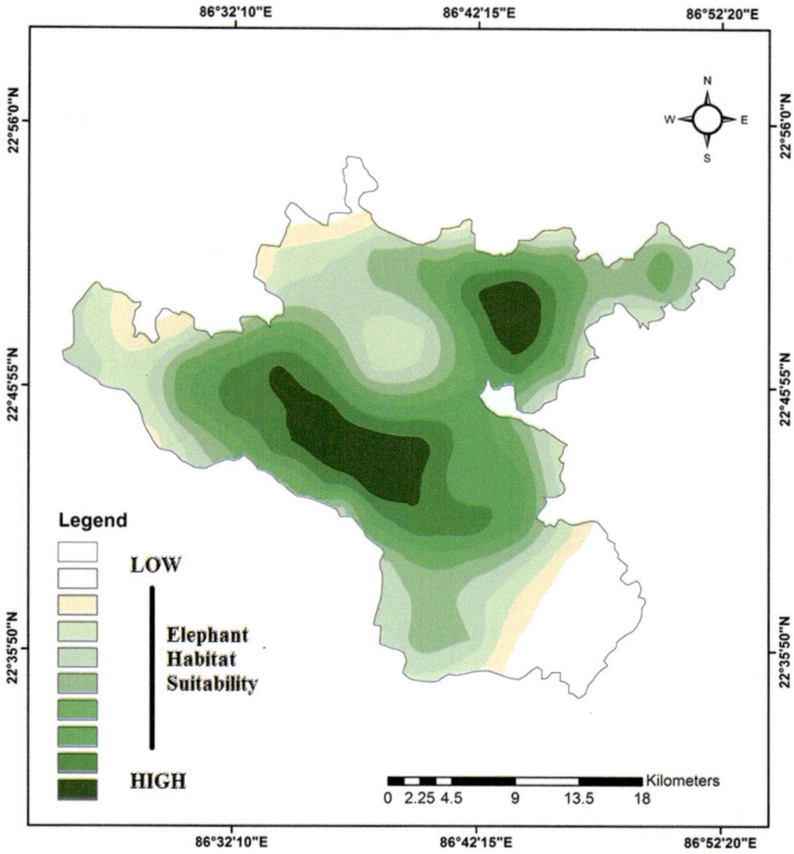

Fig. 7.19 Elephant habitat suitability zoning map of Mayurjharna Elephant Reserve on the basis of available habitat requirements

7.1.5 Potential Biodiversity Zone Demarcation

On the basis of species specific habitats, biodiversity zone demarcation is one of the ways to protect endangered species. Wildlife in their life span uses different habitats in a landscape. This method helps to understand their biological behaviour and will be used for conservation. For example elephants in Mayurjharna Elephant Reserve take rest long time in a highly suitable zone (Fig. 7.19) during their migration period (Mandal et al. 2021). Even now elephant birth incident in the core areas of the highly suitable zone also confirm the importance of this area for elephant conservation. For this, not only demarcating the boundary of such biodiversity zone but also regular investigation is needed for its ecological crisis and conflict identification. Implementation of wildlife act in these demarcated areas should be strict. These steps encourage wildlife stability and conservation as well as good and healthy biodiversity that balance regional ecosystem.

References

Abhijitha CS, Areendran G, Raj K, Bhat P, Sahana M (2021) Habitat linkages for Asian elephants in Central Indian landscape. In: Habitat, ecology and ekistics. Springer, Cham, pp 75–89

Baldus RD, Hahn R, Ellis C, DeLeon SD (2007) Connecting the world's largest elephant ranges: The Selous-Niassa corridor. In: Peace parks: conservation and conflict resolution, 109–126

Chatterjee ND (2016) Man-elephant conflict: a case study from forests in West Bengal. Springer, India

Cushman SA (2006) Effects of habitat loss and fragmentation on amphibians: a review and prospectus. Biol Cons 128(2):231–240

Cushman SA, Chase M, Griffin C (2010) Mapping landscape resistance to identify corridors and barriers for elephant movement in southern Africa. In: Spatial complexity, informatics, and wildlife conservation. Springer, Tokyo, pp 349–367
De Knegt HJ, Van Langevelde F, Skidmore AK, Delsink A, Slotow R, Henley S, Prins HH (2011) The spatial scaling of habitat selection by African elephants. J Anim Ecol 80(1):270–281
Desai AA (1991) The home range of elephants and its implications for management of the Mudumalai Wildlife Sanctuary, Tamil Nadu. J Bombay Nat Hist Soc 88(2):145–156
Desai A, Hedges S (2010) Notes from the Co-chairs IUCN/SSC Asian Elephant Specialist Group. Gajah 33(2010):3–5
Dramstad W, Olson JD, Forman RTT (1996) Landscape ecology principles in landscape architecture and land-use planning. Island Press
Fahrig L (2003) Effects of habitat fragmentation on biodiversity. Annu Rev Ecol Evol Syst 34(1):487–515
Fan Y, Lindshield S (2022) Exploring human factors of wildlife conservation along a forest gap using a participatory design-build canopy bridge. Folia Primatologica 1(aop), 1–16
Farina A (2006) Principles and methods in landscape ecology. Landscape series, Springer
Farina A (2008) Principles and methods in landscape ecology: towards a science of the landscape, vol 3. Springer Science & Business Media
Fernando P, Wikramanayake ED, Janaka HK, Jayasinghe LKA, Gunawardena M, Kotagama SW, Pastorini J (2008) Ranging behavior of the Asian elephant in Sri Lanka. Mamm Biol-Z Säugetierkunde 73(1):2–13
Forman RTT (1995) Land mosaic: the ecology of landscape and regions. Cambridge University Press, Cambridge, England
Forman RT (2014) Urban ecology: science of cities. Cambridge University Press
Geijzendorffer IR, Regan EC, Pereira HM, Brotons L, Brummitt N, Gavish Y, Walters M (2016) Bridging the gap between biodiversity data and policy reporting needs: An essential biodiversity variables perspective. J Appl Ecol 53(5):1341–1350
Harper JL et al (1977) Population biology of plants. Population biology of plants. Wales, Bangor, UK
Kraus D, Murphy S, Armitage D (2021) Ten bridges on the road to recovering Canada's endangered species. Facets 6(1):1088–1127
Li H, Wu J (2007) Landscape pattern analysis: key issues and challenges. In: Key topics in landscape ecology. Cambridge University Press
Mandal M, Chatterjee ND (2018) Quantification of habitat (forest) shape complexity through geo-spatial analysis: an ecological approach in Panchet forest division in Bankura, West Bengal. Asian J Environ Ecol 6:1–8
Mandal M, Chatterjee ND (2019) Forest core demarcation using geo-spatial techniques: a habitat management approach in Panchet Forest division, Bankura, West Bengal, India. Asian J Geogr Res 2(2):1–8
Mandal M, Chattarjee ND (2020a) Land use alteration strategy to improve forest landscape structural quality in Radhanagar forest range under Bankura district. Eurasian J For Sci 8(1):1–10
Mandal M, Chattarjee ND (2020b) Geo-statistical analysis to understand nature of forest patch shape complexity in panchet forest division under Bankura district, West Bengal. Indian J Ecol 47(1):96–101
Mandal M, Chatterjee ND (2021a) Spatial alteration of fragmented forest landscape for improving structural quality of habitat: a case study from Radhanagar Forest Range, Bankura District, West Bengal, India. Geol, Ecol, Landsc 5(4):252–259
Mandal M, Chattarjee ND (2021b) Estimation of forest ecosystem quality using GIS tool in Panchet forest division, West Bengal, India. In: Forest resources resilience and conflicts. Elsevier, pp 203–213
Mandal M, Chatterjee ND (2021c) Geospatial approach-based delineation of elephant habitat suitability zones and its consequence in Mayurjharna Elephant Reserve, India. Environ Dev Sustain 23(12):17788–17809
Mandal M, Chettarjee ND (2021d) Human-elephant conflict in Joypur forest influence areas, West Bengal, India. Gajah 54:34–36
Mouquet N, Moore JL, Loreau M (2002) Plant species richness and community productivity: why the mechanism that promotes coexistence matters. Ecol Lett 5 (1):56–65
Muruthi P (2005) Human wildlife conflict: lessons learned from AWF's African heartlands. African Wildlife Foundation Working Papers, Nairobi, 10
Naidoo R, Kilian JW, Du Preez P, Beytell P, Aschenborn O, Taylor RD, Stuart-Hill G (2018) Evaluating the effectiveness of local-and regional-scale wildlife corridors using quantitative metrics of functional connectivity. Biol Cons 217:96–103
Neupane D, Kwon Y, Risch TS, Williams AC, Johnson RL (2019) Habitat use by Asian elephants: context matters. Glob Ecol Conserv 17:e00570
Osipova L, Okello MM, Njumbi SJ, Ngene S, Western D, Hayward MW, Balkenhol N (2019) Using step-selection functions to model landscape connectivity for African elephants: accounting for variability across individuals and seasons. Anim Conserv 22(1):35–48
Sawyer SC, Epps CW, Brashares JS (2011) Placing linkages among fragmented habitats: do least-cost models reflect how animals use landscapes? J Appl Ecol 48(3):668–678

Sukumar R (1989) Ecology of the Asian elephant in southern India. I. Movement and habitat utilization patterns. J Trop Ecol 5(1):1–18

Thouless CR (1996) Home ranges and social organization of female elephants in northern Kenya. Afr J Ecol 34 (3):284–297

Tripathy BR, Liu X, Songer M, Zahoor B, Wickramasinghe WMS, Mahanta KK (2021) Analysis of landscape connectivity among the habitats of Asian elephants in Keonjhar Forest Division, India. Remote Sens 13(22):4661

8 Effective Management Methods for Wildlife Conservation in General and Elephants in Specific

Abstract

Human–elephant conflict (HEC) event is high in these forest ranges. To manage such situation and restrict animals inside the habitat, proper management is needed. The forage ground of elephant habitat is depending on suitability in terms of habitat dominance, complexity and forest ranges. Administrative boundary is not considered as it is not a boundary for wildlife. This chapter proposes landscape conversion methods to improve habitat quality for wildlife conservation in general and elephants in specific. This book exclusively demonstrated the geospatial models of Landscape management methods for better understanding the Forest Landscape Planning and Wildlife Management. In this context, this book tried the best possible efforts for the application of using models with real datasets in the case study of Mayurjharna Elephant Reserve, Panchet Forest Division, Bankura Forest, Dalma Wildlife Sanctuary and Several small forest areas to establish the fact. The application of geospatial techniques would be helpful for the students, researchers, academicians, decision makers and practitioners to using these techniques at large scale. The exceptionality of this volume is its style of presenting separate methodologies and models to validate the issue for individual chapter along with citing case studies, which will generate interests of the scientific community. These modern techniques could be facilitated the respective to prepare suitable models as a result, they could apply those models for similar cases.

8.1 Wildlife Conservation: Global Scenario

Human the most active agent in the globe modify the world ecosystem for better life. Development of civilization continuously destroyed the natural ecosystem equilibrium (Hillel 1992). Use of ecosystem resource in an irrational manner adversely effect the natural system. The most livelihood commodities as food, medicine, cloths, habitation are being obtained from biodiversity (Shiva 1998; Roe et al. 2002; Nimasow and Joshi 2017) in different ways. The anthropocentric approach of utilizing natural resources makes humans so selfish that we forget to conserve natural resources.

Wildlife conservation is a very prospective way through which conservation of the entire ecosystem can be done (Eckersley 1992; Smith and Wishnie 2000). It is the procedure that protects wildlife species as well as their habitats regarding their health, population and ecosystems. Several conservation programs continuously performing their activities in local to international level though species extinction are as usual and unstoppable (Foster 2007; Adams 2016; Crist 2022). According to WWF 200 to

M. Mandal and N. Das Chatterjee, *Geo-Spatial Analysis of Forest Landscape for Wildlife Management*, GIScience and Geo-environmental Modelling, https://doi.org/10.1007/978-3-031-33606-5_8

2000 species extinction occurred every year. IUCN reported that out of all species assessed, over 27,000 are at risk of extinction and must be under conservation. Pimm et al. (2014) estimate that current species extinction rates are about 1000 times greater due to human activities than the past extinction rate. The major threats are deforestation, habitat fragmentation and loss, overexploitation, poaching, culling, pollution and climate change.

IUCN recommended endangered species list are continuously updated. Local to international level organizations also worked and they are aware of the updated list, but success was barely achieved to implement actions to protect the species (Rohlf 1991; Treves et al. 2019). Therefore, from the late 1980s, environment-aware people began to start individual, community, and NGO sector activities for protecting and conserving endangered species (Pan 2016). These are Conservation International, Fauna and Flora International, Wild Team, Wildlife Conservation Society, Audubon Society, Traffic (conservation program), Born Free Foundation, African Wildlife Defence Force, Save Cambodia's Wildlife, Wild Earth Guardians, etc.

Elephants are the largest terrestrial mammal in the globe. As a key stone species, they play a vital role for balancing ecosystem (Bond 1994; Zhang and Wang 2003; Garibaldi and Turner 2004; Libralato et al. 2006). In spite of this, species extinct regionally from some parts of the world and endangered in their existing habitat (Lombard et al. 2001; Sukumar 2003; Chen et al. 2022). They are struggling for their survival mostly in Africa and some regions of the south Asian countries such as Sri Lanka, India, Nepal, Bhutan, Bangladesh, Thailand, China, Indonesia, Malaysia, Borneo, Sumatra, and Myanmar. Therefore, many government and non-government organization are working together to protect elephant and their ecosystem. Some reputed organizations are:

- **The Asian Elephant Specialist Group (AsESG)** is an integral part of the Species Survival Commission (SSC) of the International Union for Conservation of Nature (IUCN). The AsESG shall provide the best available scientifically grounded evidence to the abundance, distribution, and demographic status of Asian elephant populations in all 13 range states. It shall periodically compile range-wide assessments which shall be publicly disseminated through the AsESG website and other outlets.
- **Save the Elephants (STE)** is a UK registered charity based in Kenya, works to sustain elephant populations and preserve the habitats in the areas elephants are found.
- **The Elephants and Bees Project** is a part of the Save Elephants' and Human–Elephant Coexistence Program in Sagalla, Kenya and Tsavo National Park. It helps to explore the use of Beehive Fences as a natural elephant deterrent that helping the farmers and farmland.
- **African Wildlife Foundation (AWF's)** is one of the reputed foundations. There programs and conservation strategies are designed to protect the wildlife and wild lands of Africa. They ensure a more sustainable future for Africa's people.
- **The International Elephant Foundation (IEF)** they take several projects for protecting elephants from several issues like poaching, seek solutions for Human–elephant conflict, equip and train community conservationists. Their work also comprises with increase knowledge of the treatment and prevention of disease and educates people.
- **The Elephant Managers Association (EMA)** having the largest group of elephant experts for promoting welfare, husbandry and scientific research of captive elephants. They engaged and give stress in public education as well as Conservation biology of the worlds wild populations of elephants.
- **Big Life Foundation ("Big Life")** one of the major approach of is organization is to protect and sustain the wildlife and habitats in the greater Amboseli ecosystem including

remaining in East Africa. It is the first organization in East Africa to achieve coordinated cross-border operations between Kenya and Tanzania.

- **Asian Nature Conservation Foundation (ANCF) & Elephant Conservation** is actively engaged in the conservation of the Asian elephant. It has embraced to protect of entire landscapes and promotes field research projects and also they give stress on ground conservation initiatives pertaining to the Asian elephant and its habitat.
- **Save Elephant Foundation** is a Thai organization dedicated to providing care and assistance to Thailand's captive elephant population. Besides these they take a multifaceted approach involving rehabilitation programs and rescue, educational ecotourism, local community involvement and environmental conservation programs.
- **The Sri Lanka Wildlife Conservation Society (SLWCS)** is a US-based, non-profit committed to developing a sustainable model for wildlife conservation in Sri Lanka. There focus is on helping people, protect elephants and other wildlife co-exists quietly. One of the major objectives of the foundation is protecting wild elephant & biodiversity conservation.
- **EleAid—Supporting Asian Elephant Conservation** it has engaged for conservation and welfare of the Asian elephant. EleAid projects related to ensure the survival of the Asian elephant and also improve the quality of life of elephants.
- **IUCN Asian Elephant Specialist Group (AsESG)** is an integral part of the Species Survival Commission (SSC) of the International Union for Conservation of Nature (IUCN). The AsESG shall provide the best available scientifically grounded evidence to the abundance, distribution, and demographic status of Asian elephant populations in all 13 range states. It shall periodically compile range-wide assessments which shall be publicly disseminated via the AsESG website and other outlets.

8.2 Human–Elephant Conflict (HEC) Management

Different potential measures are taken in a different country to manage wildlife conflict as well as HEC. Usually, these measures are categorized in two basic types. These are offensive and defensive techniques. Defensive methods related to make obstacle or barrier through their movement path which decreases immediate contact between man and wildlife. Partially it may useful to conserve biodiversity as well as to protect human life and assets (Marcot et al. 2011). Offensive methods are concerning with the behaviour of the elephant. In this method different kinds of measures are considered for wildlife behaviours study. Indirectly these measures may restrict conflict as argued by Barnes (1996). These two types of methods are applicable through biological and mechanical way. According to Gross et al. (2022) HEC characters so different in different region therefore measures requirement quite different and it should be carried out on regional basis. They aggregate all HEC characters worldwide and simulate a comprehensive strategy for several sections for HEC management. The present chapter considered the same strategy and modified it based on regional policies, problems and local stakeholder's demands (Table 8.1).

8.2.1 Biological Measures

Biological measures have been used in wildlife conflict management from early days (Kellert et al. 1996; Van Eeden et al. 2018). These traditional methods have to use in recent problems for managing wildlife condition with some technical modifications (Treves et al. 2006). A branch of biotechnology introduces and practices different kind of methods for conserving biodiversity. Predicting wildlife conflict and management is the prerogative of this branch of science. Many such projects are significant in this respect to reduce human-animal conflict like

Table 8.1 Region based comprehensive strategies for HEC management

Conflict management strategy	Management techniques	Pattern and processes
Environment policies	International projects	Biodiversity conservation, species migration
	National scheme	Construction of sanctuary, national parks, habitat reserve through successive plan
	Local level	Problem oriented guidance and community participation
Spatial reforms and planning	Spatial population control	Corridor maintenance, managing existing animal community, scientific infrastructure development
	Landuse policy and plan	Habitat zonation, forestry like joint forest management, habitat structural quality enhancement
	Change in crop type and pattern	Concentric agriculture instead of spreading, more yields crop practice, more protection should give on valuable crops, unattractive crops cultivation in the vicinity of conflict prone areas
	Human habitation protection	Change in housing type
	Water point development	Water source inside the habitat for elephant
Social awareness	Community base	Community selection and participation, regular-based people involvement, interest growing in resource sharing, performance award for active group
	Institutional base	Promotion of different kind of workshop and training, To learn way of living with elephant
	Local base	Training for decision makers like teachers and local leaders, subject in academic syllabus from preliminary stage
Technical steps	Obstacle	Fencing, trenching
	Modern devices	Sound system, high power torches, fire burning instrument
	Bio devices	Chilli smoke and fence, beehive cultivation for fencing, solar plant for fencing
	Warning system	Mobile massaging, traffic alert signal, satellite tracking and geofencing
Economic assurance	Ex-gratia for losses	Government policy or scheme, private scheme
	Insurance	Community-based private insurance
Supervise HEC	Elephant dispersal characters	GPS, detail grass rout information, community-based monitoring, agency-based monitoring

Modified after Gross et al. (2022)

HEC in Sri Lanka, China, Thailand, Malaysia and Africa.

8.2.1.1 Crop Type and Changing Pattern

Elephants avoid a large variety of plants at a time due to chemical defences (Sukumar 2003). Therefore, altered crop means unattractive crop may be proposed as a measure to reduce crop raiding in the HEC prone areas (Naha et al. 2020; Baskaran 2013). It is common to see that paddy and horticulture field are extremely damaged by elephant in West Bengal (Mishra 1971; Desai and Hedges 2010; Mandal and Chatterjee 2020). They preferred these types of secondary vegetation as their favourite fodder. This attraction may decrease if unpalatable or unattractive crops are cultivated (Naughton-Treves 1998; Fernando et al. 2008). Some cultivation type may be used as a buffer and create obstacle for elephant for their movement across the agricultural areas. These types of crops or horticultural products may provide alternative income to the farmers. Elephants stay away from crops like Pineapple, Ginger, Sesame, Chilli, Capsicum (Parker and Osborn 2006), Colocasia, Turmeric and Lemon, etc. HEC may be minimized by cultivating these crops (Plate 8.1) near forest fringe lands. In a parallel way local government should take initiative to encourage stakeholders by implementing policies like subsidies over sheds, plants, fertilizer and all infrastructure related to cultivation processes.

8.2.1.2 Bio Fencing

Fence is used to protect animal attacks traditional rural housing settlements especially in South West Bengal. The techniques could be used to mitigate Human–elephant conflict if the fence making materials are unattractive and avoided by elephants. Some indigenous thorny plants and cactus plantation be used as a fence around the hosing colony. Another bio-measure is infusing chilli pepper on clothes and hanging them on the fences. The smoke and smell of chilli is strictly avoided by elephant as argued by Wildlife Conservation Society, 2005 in Africa. Another technique is establishing bee colonies at the borders to keep elephants away from the local communities. According to Muruthi (2005) elephants avoid spiny plant and large sharp rocks by Hanks (2006) through their movement route. Therefore, plantations by thorny plants are like cacti, bamboo, agave and bougainvillea surrounding the village, hut or crop field may be a measure to minimize HEC. Unfortunately these measures are applicable to a very small scale. Another consideration is the thickness of bio fence which is very important to protect from elephant crossing.

8.2.1.3 Increases Food Availability Inside the Forest Habitat

According to Chanda (1996) favourable and in-toxic food availability inside the forest core habitat is another process to control HEC. Avalable amount of food and perennial water source inside the habitat may restrict elephant to move in the habitat fringe areas. Though Zone-3 is a part of Mayur Jharna Elephant Reserve in spite of migratory elephants and some residential bulls do not colonize here due to unavailability of water and food after Kharif season (Fig. 8.1). Some recommended species plantation inside the forest should be a proper method to control HEC especially in elephant domain forest in West

Plate 8.1 Unattractive crops for elephants: **a** Chilli, **b** Pineapple, **c** Sesame, **d** Turmeric, **e** Colocasia and **f** Lemon

Bengal (Chatterjee and Mandal 2020). These plant species are common but favourable for elephants as food opinioned by several field expert and field biologists. Vacant or unplanted area inside the forest can be utilized as a foraging ground by planting such indigenous favourable plants like *Dhaw-Anogeissus latifolia, Bamboo, Mahwa-Madhuca indica, Kathal-Artocarpus, integrifolia, Karchmola-lanneagrandis, v, banana* and *Piasal-Pterocarpus marsupium,* etc. (Plate 8.2). These management strategies may possible when related departmental activities and people awareness may be strong (Hedges and Gunaryadi 2010).

8.2.2 Mechanical Measures

Instrumental or mechanical measures are very common and frequently used in HEC management. Installations of these measures are very expensive and related to political, social and cultural dimensions. But once implemented can give fruitful results. Some desired measures especially in South West Bengal are listed in next subsection.

8.2.2.1 Trenching Along High Risk Zone

According to Suresh (1992) only 15 km ditch played a great role to control human–elephant conflict around the Bannerghatta National Park. Ditch beside an electric fence constructed in 1997–1998 was highly helpful to control Human–elephant conflict in Sri Lanka (Jayant et al. 2007). So, trenching or ditching is a kind of landuse conversion method by which restrict movement and HEC may be managed. Frequent elephant's movement in surrounding risk villages may be diverted by artificial ditching (Fig. 8.4).

Plate 8.2 Recommended plant species for elephant forage inside the forest. **a** Bamboo, **b** Kathal, **c** Mahwa, **d** Banana, **e** Karchmola, **f** Sal, **g** Dhaw, **h** Piasal and **i** Atang

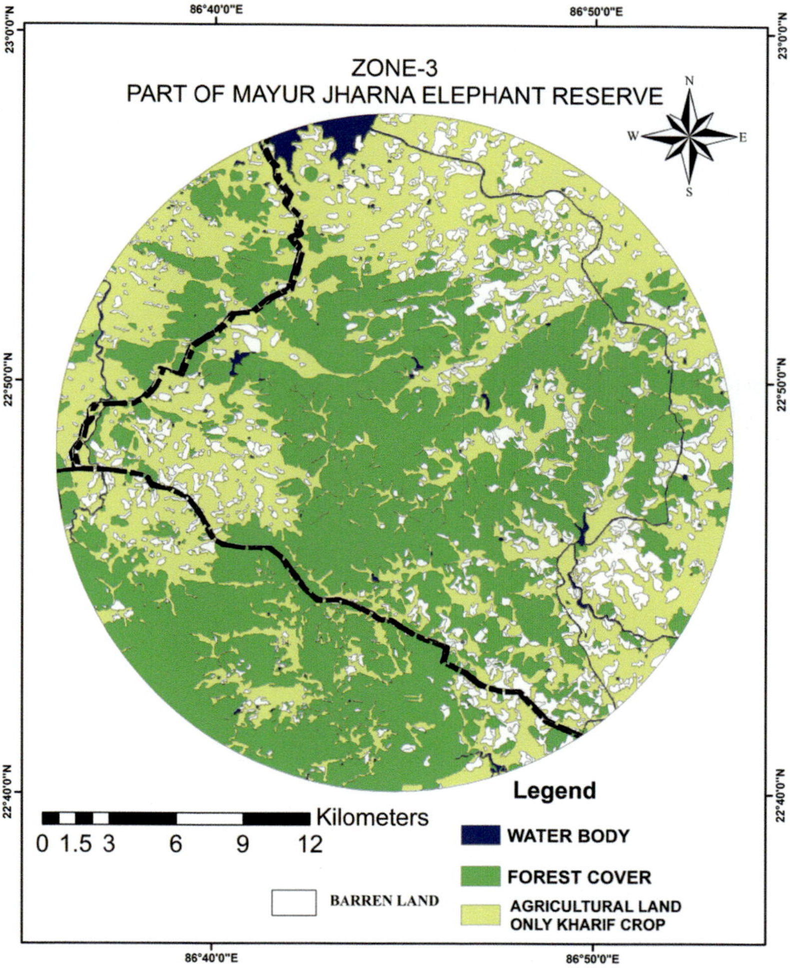

Fig. 8.1 Shows fewer perennial water sources in zone-3, part of Mayur Jharna Elephant Reserve

In high HEC region like zone-1 and zone-2 under South West Bengal, trench may reduce the frequent contact between stakeholders and elephants as well as other wildlife. Maintenance and construction of the artificial trench is very expensive and important to manage this situation. Both covered and uncovered ditches have been mostly used in African continent to restrict elephants from cultivated regions with significant success argued by Lamarque et al. (2009). Tranches around the villages may be one of the significant measures to protect the sudden conflict of wild animal in zone-1 and zone-2. It may save human life and unexpected incident like crop riding related to HEC (Plate 8.3).

8.2.2.2 Advance Alert System

Warning system is an ancient method commonly used by primitive clan to safe from wilderness. Multi-warning systems now used in man–animal conflict mitigation as well as biodiversity management (Lewis et al. 2016). Alarms system raise awareness among the farmers at the fringe of crop fields. They suddenly react to further activities to prevent crop raiding using these alarms sound (Fernando et al. 2008). Another type of alarm system is ringing bell connected with a wire. The positive results could get from different elephant domain of African continent and India also argued by several scholars (Osborn and Parker 2003; Osborn and Hill 2005;

Plate 8.3 Various Human–elephant conflict (Nature and Character) nature and characters in Rupnarayan Forest Division, in Pachim Medinipur District. Photos by Rakesh Singha Dev

Radhakrishna and Sinha 2010). Ringing the bell with connected wire is appropriate method for small-scale management. It is commonly applicable in smallest clustering village or edge between nearest crop field. Another instrumental measure is crop field fencing (Mehta et al. 2020) by beating aluminium sheet or tin. It generates sound when animal crosses or touches the fence. Villagers can aware to prevent unexpected situation when hearing this sound. Another warning instrument is used massively for warning is GPS. Exact elephant location by GPS is now used enormously to minimize HEC (Fernando et al. 2008). A networking sharing system by which all mobile consumers will get massage under influence areas about nearest elephant's exact location then unexpected incident may reduce. This information must update by hour an hour by SMS. Even tourists have to register her mobile number for getting this SMS entering into this region. Another measure is light signalling system in high risk zone (Kumar et al. 2010). This method is one type of geofencing method for controlling HEC. Tall tower or lamppost may be used for signalling in some specific locations where the movement of elephant is more frequent. Two colours are generally used in these lampposts to show the presence or absence of elephants (Fig. 8.2). Elephant presence or absence detected by automated sensor or manual operating input system. This can help to keep distance and restrict people or traffic from conflict.

Wildlife death by high speed train becomes common in both South and North Bengal. Especially elephants are more prone to such accident. Several techniques are used to protect this type of events. One of the most common method is advance sound system along railway line where demarcated corroders of elephant crosses. The unpleasant sound before coming a train will prevent to move elephant and that may reduce elephant train collision.

8.2.2.3 Rehabilitation Program

After monitoring and mapping HEC zone, it may be said that location of some village units inside the forest and outside edge of the forest are very prone to HEC. This type of scatter housing units become exposed to wild animal. Frequent contact between human and elephant become regular due to sufficient water and easy food availability

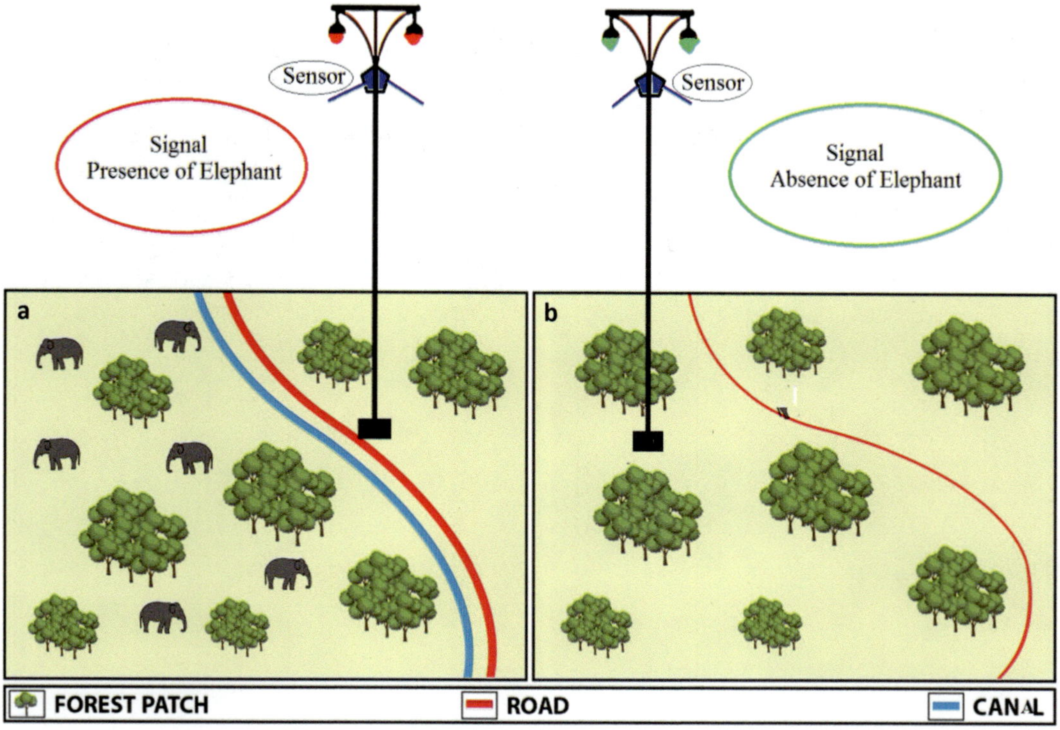

Fig. 8.2 Model for signalling information about elephant presence or absence. The signal operation may be conducted by active sensor or manual information input system

(Mandal et al. 2015) in that particular sector (Figs. 8.3 and 8.4).

It is found that scatter or doted settlement patches are exist inside the forest and even in the forest core in elephant domain areas in South West Bengal. Occurrences of these villages or isolated houses in the forest edge are more prone to conflict by wild animal, especially residential bull elephants (Mandal and Chatterjee 2021c). Therefore, shifting of these housing units from high risk zone to safe areas may be the plan to minimize accident or unexpected incident due to HEC. Rehabilitation of these houses in safe areas, i.e. outside the forest and built a colony which is surrounded by a well maintain trench is a model by which the rate of conflict may be minimized (Fig. 8.4).

8.2.2.4 Awareness

Awareness is a very fruitful and urgent method to conserve wildlife as well as biodiversity and minimize HEC (Parker et al. 2007). Nature of conflict arisen from both society and animal behaviour in a reciprocal way. Study of animal behaviour and disseminate it is the important consideration for conserving biodiversity as well as HEC management. For this reason workshop, awareness camp, drama, play, seminar, etc. needed in regular basis. But the sad thing is that village level workshop, camp, seminar, meeting, etc. are not organized institutionally or from the administration adequately in South Bengal. Forest Protection Committee can take a direct and significant role to propagate knowledge and information to the villagers.

Academic Lesson in Elementary Education

Another aspect of awareness is the inclusion of basic biodiversity module or subject in co-curriculum activates in the education system to increase awareness especially in elementary level. These lessons should be on regional basis that raise the ability and skill for protecting themselves as well as regional biodiversity. In an

Fig. 8.3 Forest edge human settlement—a suitable place for sufficient water and food for elephant

Fig. 8.4 Model for rehabilitation to built a safe colony development for minimizing HEC

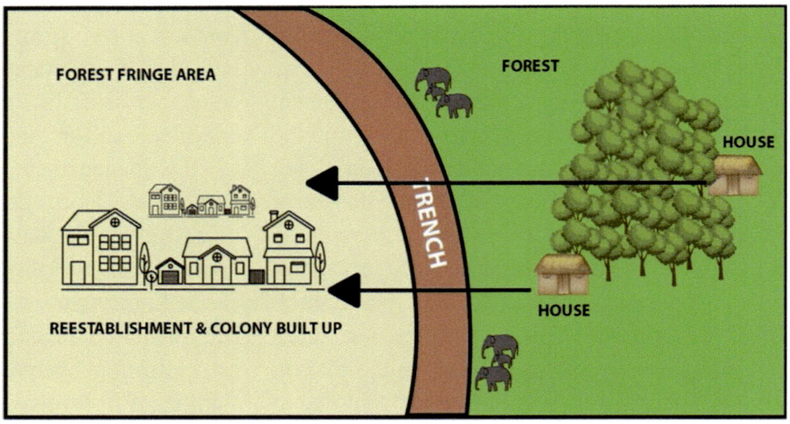

elephant domain region this academic lesson may protect stakeholders from activities like shouting, throwing stone, emotive irritation to a herd of elephant when they move.

8.2.2.5 Movement Control by 'Hula'

Expert and skilled stakeholders take a great role to drive elephant from risk zone to safe zone. These expert and skilled persons are named 'Hula' in Bengali. It is commonly found that at the time of elephants drive the young visitors disturbed elephant. They shouting loudly and throwing different objects like stone out of fun (Plate 8.4). As a result, elephant herd diverted from their route and divided into subgroups (Chatterjee 2016; Mandal and Chattarjee 2021b). This farther damage crop field extensively by trampling (Chatterjee and Chatterjee 2014;

Plate 8.4 Stakeholders views from different locations under Paschim Medinipur and Jhargram District. **a** silent moment between elephant and man, **b** irritated elephant by young man, **c** a young man ready to throw a stone and **d** member of Hula party observing elephant behaviour. Photos by Rakesh Singha Dev

Mandal and Chettarjee 2021d). It is noticed that these disturbances make elephant more irritated. As a result many times it becomes the key cause of human death as well as elephant death too. Rerouted elephants sometimes come contact with electric wire and lost their life. According to Rakes Singha Dev, a wildlife photographer in 2019 one adult male and a female elephant found dead. On 11 January night the elephants came contact with high voltage electric wires. The poor elephants died on the spot after being electrocuted in Pachim Medinipur district. Therefore, at the time of elephant driving, silence and help to Hula party to drive them is the proper way to reduce HEC. If elephant driving is completed in a proper way then the rate of crop and assets damage may be minimized (Mandal and Chatterjee 2021a). Every forest ranges of high conflict zone must have Hula party to drive and control elephant's movement. They have to train an regular basis and to take them under government or private insurance because of their life risk.

Box 8.1: Elephant Death by Electrification in Moupal Beat, Bhadutala Forest Range, Paschim Medinipur

Another female elephant died from electrocution at Salboni Block in Paschim Midnapur district, West Bengal. The local villagers said around 8 p.m. to 8:30 p.m. a herd of around 70–80 elephants had come near Jorakusma village under Moupal beat. The herd of jumbos enjoyed their food in the local paddy fields behind the Jorakusma-Susnibari primary school. At that time someone of that herd pushes an electric poll in the paddy field and an elephant came contact with high voltage electric wires. The poor elephant died immediately after being electrocution. That is the second incident took place in the same spot within one year.

Photo: Singha Dev.
Moupal Beat, Bhadutala Forest Range.

8.2.2.6 Stakeholder Participation in Biodiversity Conservation

The best way of wildlife conservation is stakeholder's participation. This concept may be useful for biodiversity management. Local people should have the rational opinions how wildlife be managed in better ways. Therefore administration and academician should try to approach the locals for saving the biodiversity by inspiring them through eractive workshops, seminars, training and biodiversity related employments. If the locals understand the importance of their local biodiversity, they will try to save them.

Such kinds of approaches were conducted by JFM in South Bengal. By this holistic operation a new model was established for forest conservation as well as wildlife in this area. The indigenous people associated with this program and restricts themselves partially from such kind of ritualistic killing of wildlife. Hunting festivals was a recognized event for endangered wildlife in this region (O'Malley 1908). But at present this festivals lost its grandeur due to transformation of culture and awareness programs. It is interesting to see that the locals sometimes protest against forest clearing, help to rescue wild animal, freely engaged for plantation and several kinds of social programs for improving biodiversity in their surroundings. During survey we come to know about some traditional strategies (Plate 8.5) that helps to modify the models for managing HWC as well as conserving biodiversity in this region.

Plate 8.5 During survey time **a** a tribal woman in the Arabari forest waiting for forest product collection, **b** native dwellers carry the Sal tree leaves, **c** opinion sharing from a Forest Beat In-charge at Chhatna Forest Range and **d** interaction of a Forest Range Officer at Arabari forest

Box 8.2: Ritualistic Hunting is Possibly One of the Biggest Killers of Wildlife in South West Bengal

Ritualistic hunting is also a part of the communities of this region. Indiscriminate killing of thousands of protected wildlife in rural Bengal, in the name of 'ritualistic hunting' happens in this region. Ritualistic hunting is possibly one of the biggest killers of wildlife in South West Bengal, threatened species, including elephant, pangolin, fishing cat, hyena, wild boar and wolf, to name a few. The problem is most dominant in the southern districts of Purulia, Jhargram, Bankura and Paschim Medinipur. Armed with bows and arrows, axes, spades, swords, knives, hammers, sling shots, nets, traps and other sharp weapons, hunters' groups flock to various hunting locations and set out to kill every wild animal they can find. It follows the brutal slaughter of thousands of mammals, birds and reptiles, all protected under the Wildlife (Protection) Act, 1972. Species killed during these 'hunt fests' include fishing cats, monitor lizards, pangolins, jungle cats, porcupines, jackals, Bengal foxes, wolves, wild boars, civets and birds like pittas, owls, barbets, francolins, bitterns and jacanas, to name just a few. Even a tiger that had appeared in Lalgarh in Jhargram in 2018 was hunted down by tribal hunters also. Some of the hunters also sell off the skin and body parts of the poached animals to cater to the illegal trade in wildlife.

With so many laws and prohibition unfortunately, this hunting festivals cannot be stooped permanently and the wildlife in this region are affected highly. However, awareness programs and alternate income from the wildlife that has been introduced in the area by the forest department has lowered the incident of mass killing of wildlife to some extent. Now the hunters also feel the importance of these animals in their regional biodiversity.

Ecotourism development may be the useful to ascertain the livelihood of the local community. In Ecotourism, stakeholders participation helps to achieve the goals of inclusive management. By this approach the locals benefited economically that ensures ecological sustainability. Another way of participation of local people is certification of training, workshop, seminars, competition, etc. that indirectly helps the people for gathering knowledge about wildlife as well as biodiversity. Ultimately, biodiversity may be conserve in proper way when academic, cultural, social, administration, economic and political systems are working together.

References

Adams M (2016) Ecological crisis, sustainability and the psychosocial subject. Palgrave Macmillan, London

Barnes RF (1996) The conflict between humans and elephants in the central African forests. Mammal Rev 26(2–3):67–80

Baskaran N (2013) CEPF Western Ghats special series: an overview of Asian Elephants in the Western Ghats, southern India: implications for the conservation of Western Ghats ecology. J Threatened Taxa 5 (14):4854–4870

Bond WJ (1994) Keystone species. In: Biodiversity and ecosystem function. Springer, Berlin, Heidelberg, pp 237–253

Chanda S (1996) Man-elephant conflict in South West Bengal. Joint Director of Sundarban Biosphere Reserve, West Bengal) IFS (1996 Batch), 1–8

Chatterjee ND (2016) Man-elephant conflict: a case study from forests in West Bengal India. Springer International Publishing

Chatterjee ND, Chatterjee S (2014) Changing habitat and elephant migration from Dalma Wildlife Sanctuary, Jharkhand to Panchet Forest Division, Bankura, West Bengal: a biogeographical analysis. In Climate change and biodiversity. Springer, Tokyo, pp 209–222

Chatterjee ND, Mandal M (2020) Human-elephant confect in Panchet Forest Division, Bankura, West Bengal, India. Gajah 51:10–15

Chen Y, Sun Y, Atzeni L, Gibson L, Hua M, Li K, Dudgeon D (2022) Anthropogenic pressures increase extinction risk of an isolated Asian elephant (Elephas maximus) population in southwestern China, as revealed by a combination of molecular-and landscape-scale approaches. Integr Zoolgy

Crist E (2022) Witnessing mass extinction: what's invisible, what's visible, what's possible. Biol Cons 275:109696

Desai AA, Hedges S (2010) Notes from the Co-chairs IUCN/SSC Asian Elephant Specialist Group. Gajah 33:3–6

Eckersley R (1992) Environmentalism and political theory: toward an ecocentric approach. Suny Press

Fernando P, Wikramanayake ED, Janaka HK, Jayasinghe LKA, Gunawardena M, Kotagama SW, Pastorini J (2008) Ranging behavior of the Asian elephant in Sri Lanka. Mamm Biol-Z Säugetierkunde 73(1):2–13

Foster JB (2007) The ecology of destruction. Mon Rev 58 (9):1

Garibaldi A, Turner N (2004) Cultural keystone species: implications for ecological conservation and restoration. Ecol Soc 9(3)

Gross EM, Pereira JG, Shaba T, Bilério S, Kumchedwa B, Lienenlüke S (2022) Exploring routes to coexistence: developing and testing a human-elephant conflict-management framework for African elephant-range countries. Diversity 14(7):525

Hanks J (2006) Mitigation of human–elephant conflict in the Kavango-Zambezi transfrontier conservation area through community based problem animal control, with particular reference to the use of Chilli Peppers. Final Report. December

Hedges S, Gunaryadi D (2010) Reducing human–elephant conflict: do chillies help deter elephants from entering crop fields? Oryx 44(1):139–146

Hillel D (1992) Out of the Earth: civilization and the life of the soil. Univ of California Press

Jayant K, Mehta P, Boominathan D, Chaudhuri S (2007) A study of man-elephant conflict in Nagarhole National Park and surrounding areas of Kodagu district in Karnataka, India. Final Report. Envirosearch

Kellert SR, Black M, Rush CR, Bath AJ (1996) Human culture and large carnivore conservation in North America. Conserv Biol 10(4):977–990

Kumar MA, Mudappa D, Raman TRS (2010) Asian elephant Elephas maximus habitat use and ranging in fragmented rainforest and plantations in the Anamalai Hills, India. Trop Conserv Sci 3(2):143–158

Lamarque F, Anderson J, Fergusson R, Lagrange M, Osei-Owusu Y, Bakker L et al (2009) Human-wildlife conflict in Africa: causes, consequences and management strategies. Food and Agriculture Organization of the United Nations (FAO)

Lewis AL, Baird TD, Sorice MG (2016) Mobile phone use and human–wildlife conflict in northern Tanzania. Environ Manage 58(1):117–129

Libralato S, Christensen V, Pauly D (2006) A method for identifying keystone species in food web models. Ecol Model 195(3–4):153–171

Lombard AT, Johnson CF, Cowling RM, Pressey RL (2001) Protecting plants from elephants: botanical reserve scenarios within the Addo Elephant National Park, South Africa. Biol Cons 102(2):191–203

Mandal M, Chatterjee ND (2020) Elephant's habitat suitability assessment through geo spatial quantification in Panchet forest division, West Bengal. Ecofeminism Clim Change

Mandal M, Chatterjee ND (2021a) Geospatial approach-based delineation of elephant habitat suitability zones and its consequence in Mayurjharna Elephant Reserve, India. Environ Dev Sustain 23(12):17788–17809

Mandal M, Chattarjee ND (2021b) Estimation of forest ecosystem quality using GIS tool in Panchet forest division, West Bengal, India. In: Forest resources resilience and conflicts. Elsevier, pp 203–213

Mandal M, Chatterjee ND (2021c) Forest landscape and its ecological quality: a stepwise spatiotemporal evaluation through patch-matrix model in Jhargram District, West Bengal State, India. Reg Sustain 2 (2):164–176

Mandal M, Chettarjee ND (2021d) Human-elephant conflict in Joypur Forest influence areas, West Bengal, India. Gajah 54:34–36

Mandal M, Chettarjee ND, Hazra J (2015) Elephant migration and colonization in Bankura district, West Bengal, India. Vidyasagar University. Indian J Geogr Environ 14:46–52

Marcot BG, Kumar A, Talukdar G, Srivastava AK (2011) Habitat relationships of Asian elephants in shifting-cultivation landscapes of Meghalaya, northeast India. Gahah 34:8–17

Mehta PM, Kulkarni J, Athale U (2020) There's many a way to keep the elephant away. A review of crop protection and elephant management techniques. Wildlife Research and Conservation Society. India

Mishra J (1971) Assessment of annual damage to crops by elephants in Palamau District, Bihar. Bombay Natur Hist Soc J

Muruthi P (2005) Human wildlife conflict: lessons learned from AWF's African heartlands. African Wildlife Foundation Working Papers, Nairobi, 10

Naha D, Dash SK, Chettri A, Roy A, Sathyakumar S (2020) Elephants in the neighborhood: patterns of crop-raiding by Asian elephants within a fragmented landscape of Eastern India. PeerJ 8:e9399

Naughton-Treves L (1998) Predicting patterns of crop damage by wildlife around Kibale National Park, Uganda. Conserv Biol 12(1):156–168

Nimasow G, Joshi RC (2017) Changing identity, livelihood and biodiversity of Indigenous communities in The Eastern Himalaya with special reference to Aka tribe. In: Societies, social inequalities and marginalization. Springer, Cham, pp 145–160

O'Malley LSS (1908) Bengal District Gazetteers: Bankura. Bengal Secretariat Book Depot, Calcutta

Osborn FV, Hill CM (2005) Techniques to reduce crop loss: human and. People and Wildlife, Conflict or Co-existence? 72

Osborn FV, Parker GE (2003) Towards an integrated approach for reducing the conflict between elephants and people: a review of current research. Oryx 37 (1):80–84

Pan J (2016) China's environmental governing and ecological civilization. Springer, Berlin Heidelberg

Parker GE, Osborn FV (2006) Investigating the potential for chilli Capsicum spp. to reduce human-wildlife conflict in Zimbabwe. Oryx 40(3):343–346

Parker GE, Osborn FV, Hoarse RE (2007) Human-elephant conflict mitigation: a training course for community-based approaches in Africa (participant's manual)

Pimm SL, Jenkins CN, Abell R, Brooks TM, Gittleman JL, Joppa LN, Sexton JO (2014) The biodiversity of species and their rates of extinction, distribution, and protection. Science 344(6187):1246752

Radhakrishna S, Sinha A (2010) Living with elephants: Exploring the nature and cause of human-elephant conflict in India (NIAS Backgrounder No. B2-2010) (No. B2-201). NIAS

Roe D, Mulliken T, Milledge S, Mremi J, Mosha S, Grieg-Gran M (2002) Making a killing or making a living? Wildlife trade, trade controls and rural livelihoods. Biodiversity and livelihoods issues no 6. IIED, London, UK

Rohlf DJ (1991) Six biological reasons why the Endangered Species Act doesn't work—and what to do about it. Conserv Biol 5(3):273–282

Shiva MP (1998) Inventory of forest resources for sustainable management & biodiversity conservation with lists of multipurpose tree species yielding both timber & non-timber forest products (NTFPs), and shrub & herb species of NTFP importance. Indus Publishing

Smith EA, Wishnie M (2000) Conservation and subsistence in small-scale societies. Ann Rev Anthropol 493–524

Sukumar R (2003) The living elephants: evolutionary ecology, behaviour, and conservation. Oxford University Press

Suresh M (1992) Economics of elephant proof trenches and electric fences in Bannerughatta National Park. Dissertation. Submitted to the Wildlife Institute of India. Dehradun

Treves A, Wallace RB, Naughton-Treves L, Morales A (2006) Co-managing human–wildlife conflicts: a review. Hum Dimens Wildl 11(6):383–396

Treves A, Santiago-Ávila FJ, Lynn WS (2019) Just preservation. Biol Cons 229:134–141

Van Eeden LM, Crowther MS, Dickman CR, Macdonald DW, Ripple WJ, Ritchie EG, Newsome TM (2018) Managing conflict between large carnivores and livestock. Conserv Biol 32(1):26–34

Zhang L, Wang N (2003) An initial study on habitat conservation of Asian elephant (Elephas maximus), with a focus on human elephant conflict in Simao, China. Biol Cons 112(3):453–459

Index

M. Mandal and N. Das Chatterjee, *Geo-Spatial Analysis of Forest Landscape for Wildlife Management*, GIScience and Geo-environmental Modelling, https://doi.org/10.1007/978-3-031-33606-5

Printed by Printforce, the Netherlands